Redrawing Humanity's Map

Kira Virginia Vinke

Redrawing Humanity's Map

How Climate Change Is Shaping Migration

Springer

Kira Virginia Vinke
Center for Climate and Foreign Policy
German Council on Foreign Relations
Berlin, Germany

ISBN 978-3-031-85986-1   ISBN 978-3-031-85987-8   (eBook)
https://doi.org/10.1007/978-3-031-85987-8

Translation from the German language edition: "Sturmnomaden: Wie der Klimawandel unsere Häuser zerstört" by Kira Virginia Vinke, © dtv Verlagsgesellschaft 2022. Published by dtv Verlagsgesellschaft. All Rights Reserved.

Cover Photo: © Manolo Ty

Maps: © Peter Palm, Berlin, Germany

© The Editor(s) (if applicable) and The Author(s), under exclusive license to Springer Nature Switzerland AG 2025

This work is subject to copyright. All rights are solely and exclusively licensed by the Publisher, whether the whole or part of the material is concerned, specifically the rights of translation, reprinting, reuse of illustrations, recitation, broadcasting, reproduction on microfilms or in any other physical way, and transmission or information storage and retrieval, electronic adaptation, computer software, or by similar or dissimilar methodology now known or hereafter developed.
The use of general descriptive names, registered names, trademarks, service marks, etc. in this publication does not imply, even in the absence of a specific statement, that such names are exempt from the relevant protective laws and regulations and therefore free for general use.
The publisher, the authors and the editors are safe to assume that the advice and information in this book are believed to be true and accurate at the date of publication. Neither the publisher nor the authors or the editors give a warranty, expressed or implied, with respect to the material contained herein or for any errors or omissions that may have been made. The publisher remains neutral with regard to jurisdictional claims in published maps and institutional affiliations.

This Springer imprint is published by the registered company Springer Nature Switzerland AG
The registered company address is: Gewerbestrasse 11, 6330 Cham, Switzerland

If disposing of this product, please recycle the paper.

Preface

Migration Narratives

In the middle of the night, they must leave. Together with the children, she departs. Aware of the possibility of failure, fear and doubt accompany her risky journey. But she risks her own life and the lives of her daughters and sons to live in safety.

Whose story of migration is this?

The Ukrainian war refugee? A Syrian family story? The attempt to escape the vicious circle of poverty and violence in Burkina Faso? The evacuation from a coastline hit by a superstorm in Bangladesh? Your own?

Currently, there are about 120 million ways to answer this question. Because more people than ever are forcibly displaced, more than 120 million, as the United Nations High Commissioner for Refugees (UNHCR) reported in 2024.

Historically, too, there are numerous stories of migration that continue to shape our identity and values today. Displacement and migration connect us as people rather than divide us—even if the categories "refugee" or "migrant" suggest otherwise. Although migration and displacement are part of our historical DNA, many look with horror at the events in the Mediterranean, at the border between the USA and Mexico, or in the Sahara and ask themselves what drives people to expose themselves—and not infrequently also their children—to such extreme risks. One image that no one who has seen it is likely to forget is that of the body of little Alan Kurdi. The Syrian boy, barely three years old, died while fleeing across the Mediterranean Sea in 2015, his lifeless body washed up on a Turkish beach and recovered by a police officer. What happens in the countries of origin that people are willing to leave behind everything for an uncertain future? It is important to analyze this carefully, because the reasons for migration are multifold.

In 2022, the incomprehensible happens. Russia, under the leadership of Vladimir Putin, attacks Ukraine. Bombs fall on the capital Kiev, the port city of Mariupol, Kharkiv, and other cities. Lviv, near the Polish border, is also bombed. Millions flee the country, and millions more become internally displaced. How their lives will continue, and whether they will be able to return after the fighting ends or will have to settle permanently in foreign lands is an open question at the time of this book's publication.

The fact that in addition to armed conflicts, from which we can hardly detach our attention, other factors also force people to flee and migrate is shown by the report of the German Federal Government's Commission on the Root Causes of Displacement "Preventing Crises, Creating Perspectives, Protecting People," published in 2021. The commission concludes that only rarely—as in the case of Ukraine—does a single cause lead to the migration of a person. Rather, the pressure usually arises from a web of direct drivers, such as war, persecution, violence, and poverty, as well as indirect drivers. The latter include climate change, whose devastating consequences destroy livelihoods and expose people to enormous risks.

This book, which combines eight years of research, is about the impact of climate extremes on migration and displacement. I address not only immediate environmental changes but also how these changes are reflected in our social systems and in what ways governments can promote or impede adaptation to climate change. In numerous countries, I have spoken with people who have themselves had to leave their ancestral homelands due to climate impacts, with people who are trying to support them, and also with those who want to prevent migration. These interviews have produced a compendium of scientific work but also a complex overall picture of the state of our planet and the people who live on it.

I would like to share this mosaic made up of many individual pieces with you and in doing so amplify the voices of people who have already entered the battle with the forces of nature. Many of them have become wanderers whose homes have been destroyed and whose future is uncertain—in part because increasing storms, droughts, and floods could drive them away repeatedly. Their testimony holds powerful warnings for us. For without a stable world climate, there can be no human development. Without climate protection, there can be no peace. Without consistent change, there can be no hope. Only a rapid phase-out of fossil fuels and a transformation to sustainability in all sectors can still limit the damage. Already, many ecosystems are stressed by decades of political climate lethargy, causing people to consider moving to greener pastures. What scenario awaits us in the coming decades? Migration is part of our history, the history of humanity. We will decide on the next chapter.

About the Structure of the Book

The first chapter is about the specific migration patterns: migration within one's own country or across borders, voluntary or forced, seasonal or permanent migration, and the question of the extent to which climate change affects our social system. In Chap. 2 the legal frameworks and questions of protection are addressed. The following chapters cover different regional contexts. Chapter 3 assesses climate impacts threatening the livelihoods in small island states as a result of sea level rise; Chap. 4 the violent conflicts against the backdrop of hydro-climatic extremes in the

Sahel; Chap. 5 the superstorms in the Philippines and Bangladesh; Chap. 6 the destruction of biodiversity and the multiple consequences of climate change in the Amazon rainforest; and Chap. 7 the challenges facing Germany and Switzerland. This geographically focused analysis is followed by a concrete proposal on how to support particularly threatened individuals (Chap. 8), as well as fragments of hope that point to ways out of climate chaos with innovations, best practices, and civil society initiatives (Chap. 9).

Methodological Note

All interview data which is cited here has been obtained with prior written informed consent and the research design of each field trip was reviewed by a research and/or ethics committee. The names of interviewees have largely been changed for privacy reasons, as indicated in the footnotes.

Berlin, Germany											Kira Virginia Vinke

Contents

1	**Departure into the Unknown: Climate Migration in the Twenty-First Century**.	1
	Internally Displaced	3
	Migration Patterns in Times of Climate Change.	4
	Decades of Climate Migration	5
	Small Data, Big Data, No Data?.	6
	The Nomadization of Our Way of Life.	8
	What Climate Migration Has to Do with Us	9
	Caught Up in Climate Chaos	10
	Corona and the Climate	11
	Future Prospects	11
	Thesis	13
2	**The Right to Stay and the Freedom to Go: Legal Aspects of Climate Migration**	15
	The Refugee Convention and its Limitations	16
	The Principle of Non-refoulement	17
	Climate Displacement within Countries.	19
	Holes in the Protections Afforded by the Universal Declaration of Human Rights.	21
	Climate Lawsuits and the Human Right to a Clean Environment	21
	Ecocide: A Crime against Peace?.	24
	State Without Territory?	26
	Thesis	28
3	**Islands Without a Future? The Disappearing Paradise of the Small Island States**.	29
	Desolate Homeland: Survival in Paradise.	30
	A Deck Full of Blood	33
	Superyachts.	35
	Atolls in Between World Powers	35
	The Expulsion from Bikini and the Nuclear Legacy of the US	37

	One Point Five to Stay Alive....................................	39
	Planned Relocations in Fiji......................................	43
	In the Eye of Irma: Superstorms in the Caribbean	45
	Evacuation by Force of Arms	48
	Billionaires and Homelessness: Who Owns the Islands?............	49
	Toxic Algae Belt	50
	Grab-and-Go Capitalism...	51
	Crises: Welcome to the Anthropocene	52
	Thesis	53
4	**Conflicts Between Nomadic Herders and Sedentary Farmers in the Sahel**...	55
	Europe's External Borders in the Desert.........................	58
	Burkina Faso: Terror Under the Starry Sky	63
	Successful Farmers at the Limit	64
	Ethiopia: A Country in Transition...............................	67
	A Mighty Dam	71
	Sahel at Crossroads...	72
	Thesis	72
5	**Superstorms: Long-Term Impacts in the Philippines and Bangladesh**...	73
	Hurricanes, Cyclones, and Typhoons	74
	Climate Diplomats in Tears......................................	75
	Unsafe Shelters: Typhoons Destroy the Philippines................	76
	New Climate, New Challenges...................................	78
	Climate Crisis Hotspot Bangladesh	80
	Worse Extreme Events, Fewer Fatalities.........................	81
	Megacity Dhaka in the Grip of the Forces of Nature	82
	Climate Impacts in Slums.......................................	82
	Taking on Debt to Survive	86
	Tiger Conservation Versus Human Development..................	87
	Cities for Climate Migrants?.....................................	91
	One-Third of National Territory under Water....................	93
	Thesis	94
6	**Fire in the Rainforest: Biodiversity Crisis in the Amazon Basin**	95
	Brazil: Guardian of the Amazon Basin...........................	96
	Tireless Activists...	99
	Lula III: New Times, Old Challenges............................	100
	Climate Change and Species Loss	102
	The Role of Diversity	103
	Planetary Health	104
	Peru: Three Vegetation Zones and Many Challenges	105
	Lima: City Without Water......................................	107

The Ruthless Boy: El Niño	108
Weather Services at Crossroads	110
Wandering Trees and Fortune Seekers	112
The Future of Our Climate Niche	114
Thesis	114

7 Climate Crisis in Germany and Switzerland: From the Halligs to the Alps 115

Reconstruction in the Ahr Valley	117
Unheeded Warnings	119
Fossil Disaster Management	121
From the Frying Pan into the Fire: Extreme Precipitation and Climate Change	122
Displaced by Coal Mines	123
Hot Times	123
Threatened Alpine Idyll	124
German Coastlines Under Pressure	127
Near and Far Effects of Climate Change	128
Climate Migration to Europe?	129
Thesis	133

8 A Climate Passport for Climate Migrants? The Political Toolbox for the Systemic Crisis 135

Uninhabitable Lands	136
Return Impossible?	137
Prepared for Emergencies	138
Nansen's Legacy	139
A Protection Agenda	140
Climate Migration Mainstreaming	142
The Nansen Passport for Climate Displaced Persons	143
Castles in the Air and Walls on the Ground	145
Debate in the German Bundestag	146
Voices from the Small Island States	147
We Will Stay	148
Inhumane Border Policies	149
Thesis	150

9 Pathways Out of the Crisis: Fragments of Hope 151

What Climate Migration Has to Do with Racism	151
What Climate Migration Has to Do with Sexism	154
Gender Equality in Industrialized Countries	155
A Social Problem	158
Solutions Outside the Box	158
Architects for the Poorest	159
Floating Cities as a Lifeline?	160

    The Imperfect City of Tomorrow 161
    Climate Knowledge and Culture............................... 162
    Spring Heat Waves 162
    What Is Next? Judgment, Conviction, and Empathy 165

Acknowledgments 167

Bibliography 169

Chapter 1
Departure into the Unknown: Climate Migration in the Twenty-First Century

Internally Displaced ♦ Migration Patterns in Times of Climate Change ♦ Decades of Climate Migration ♦ Small Data, Big Data, No Data? ♦ The Nomadization of our Way of Life ♦ What Climate Migration Has to Do With Us ♦ Caught Up in Climate Chaos ♦ Corona and the Climate ♦ Future Prospects

On an oppressively hot day in New Delhi, while the asphalt melted outside and crosswalks became snake-like watercolors, I discussed the situation of migrant workers in India with my roommate in a run-down guesthouse. Millions of India's internal migrants toil in boom cities, mostly under the extreme conditions of working in the informal sector. They arrive from rural areas, where many were landless farmers. In the cities, without the benefit of occupational health and safety standards, they serve as the shadow engine of an increase in prosperity for the wealthy and middle class with their physically demanding labor. On some summer days, the heat is unbearable, and air pollution makes one choke, but there is no halt to work on the construction sites which are full of people who have to earn some kind of wage. I wondered: What exactly drove these migrants to flock to the cities? Was it truly in pursuit of a better life through economic advancement as the classic narrative of urbanization suggested? Were they trying to escape the downward spiral that scarcity of resources and rural poverty thrust upon them? Or was it necessity rather than choice driving them to leave their homes? Could a less visible reality of climate change be forcing their migration? At the time, in 2013, I was doing research in Delhi at The Energy and Resources Institute (Teri) University on water security and transnational river management. Little did I know that as I sipped my chai and dabbed my damp forehead that this was the beginning of my journey into understanding the very real threats of climate displacement.

The term "climate migration" generates over 15,800 results when input into the Ecosia search engine. No wonder: The fact that there is a connection between "climate" and "migration" has set off many scientific and media debates in recent years. Climate migration is a topic that attracts attention. Often, the focus is on theoretical questions of definition and projections of future migration flows. Hidden behind the objections over the sheer number of migrants to be absorbed into a community is the darker question at hand: "Will allowing these people a seat at the table mean I have to give up my prosperity, voice or privileges?"

Relatively little attention has been paid so far to those people who are already threatened and migrating today as global warming is approaching 1.5 °C above pre-industrial temperature levels. This is in part due to the complex mix of factors involved. Migration—like most human actions—is multicausal. That is to say that people migrate for reasons other than climate change. They may be in search of better job opportunities, escaping an area that is pressured by high population density or in order to support obligations to family. Climate impacts clearly influence certain drivers of migration [1]. For example, extreme weather patterns that lead to the destruction of crops or heat waves that make necessary outdoor labor unbearable.

So far, climate change has not been the dominant factor in global migratory movements. However, in 2021, the Intergovernmental Panel on Climate Change (IPCC) made clear in its assessment report that we will face more severe climate impacts across all emissions scenarios. Warming has already temporarily crossed the 1.5 °C threshold, the lower warming boundary of the Paris Climate Agreement. The realization is painful: There is no going back for the younger generation to the climate in which their grandparents grew up. The Paris Agreement is a milestone of international climate diplomacy as it outlines common targets, such as limiting global warming to certain temperature thresholds, for all member states of the United Nations Framework Convention on Climate Change. But, even if countries were to rapidly phase out fossil fuels and transform major sectors like agriculture and transportation to adhere to the agreement, temperatures would continue to rise for a period of time before a stabilization would occur. This is explained by both the inertia of our global economic system—which by no means can be decarbonized [2] overnight—and to how the Earth system responds to the changing carbon dioxide (CO_2) levels in the atmosphere. Humankind missed the opportunity to counteract warming at an early stage a few climate negotiations ago. This means that climate impacts, which have so far been a minor noise in the cacophony of factors influencing migration decisions and routes, could become a much bigger factor in the future. However, if proper urgent action is taken, it is still in mankind's hands to avert the worst climate risk projections that threaten civilization.

My research has taken me to climate change hotspot regions in South Asia, Latin America, sub-Saharan Africa, and the Pacific. These travels allowed me to witness how profound the changes already are and to see firsthand the consequences of migration for migrant people themselves and also sending and receiving communities [3]. Behind steep temperature curves and complex graphs lie the fates of individuals and communities. Many of them are underrepresented, and the enormous challenges they confront have barely found their way into discussions about climate protection plans at the national and international level. But this is not a problem only in disadvantaged nations. Even in industrialized countries like the United States and in Europe, conversations with farmers and fishermen quickly reveal that a major current is in motion, even if the surface seems calm.

The climate system is increasingly out of balance [4]. Forest fires, storm surges, and crop failure are now dominating the news. After extremely hot summers, devastating floods, and deadly droughts, climate change has arrived across the geographies. One school of thought argues: "Humans have always migrated to adapt to change." While that is true, the change that is currently occurring is dramatic and unprecedented. Sea levels are rising faster than at any time in the last 3000 years [5]. The CO_2 concentration in the atmosphere is higher than it has been in at least the last two million years. This means we are the first humans to have to contend with such a high CO_2 concentration in the atmosphere. *Homo sapiens* has only been around for about 300,000 years. We are thus in the midst of a global experiment that we ourselves started, and of which the outcome is uncertain. At the same time, the world's population has reached an all-time high. Unlike in the past, many areas are now densely populated, and resources are dwindling. Migration in the context of climate change is thus a challenge, necessity, and opportunity for humanity.

Internally Displaced

The escape route of people who leave their homelands behind to build a new existence elsewhere largely occurs within countries. Those who cross national boundaries go to nearby countries. Data from the United Nations Refugee Agency (UNHCR) [6] shows that 86% of all refugee movements go to developing and emerging economies. Many seek shelter in neighboring countries of their home country. This finding is also confirmed by the meta-analysis of my colleague Dr. Roman Hoffmann, in which he summarized and statistically evaluated the results of a large number of studies on environmental migration [7]. According to this body of work, only very few people are drawn to faraway places, such as Europe or North America. Most of them want to stay close to their respective hometowns and cultural heritage.

In this context, Professor Hans Joachim Schellnhuber, founder of the Potsdam Institute for Climate Impact Research, debunks the much used water metaphor of the "wave of refugees" or the "flow of refugees" to Europe in his book *Self-Combustion*. "The great migration is more like a trickle through the poorer regions of the world, with entire ethnic groups occasionally coming to a dead end," he writes in his comprehensive work [8]. For others, it is not even possible to migrate over longer distances. They lack the financial resources or simply the access to transportation—but more on that later.

The Internal Displacement Monitoring Centre (IDMC), which collects, processes, and analyzes data on internal displacement, recorded high numbers in 2023: 46.9 million new internal displacements [9]. This sad statistic is only indirectly related to the consequences of climate change, as a large proportion (26.4 million) of people fled natural disasters, while 20.5 million sought refuge from armed conflict. The natural disasters were predominantly weather-related. While by no means all extreme weather events are attributable to climate impacts, climate change is causing such events to occur with increasing frequency and intensity. Moreover, with population growth, an increasing number of people is living in exposed areas. Of particular concern is that in many areas, even minor floods, storms, or droughts lead to catastrophic consequences for the population because of fragile infrastructure. Those hardest hit are people who have not hit the jackpot in the "lottery of birth" to begin with. They live in poverty—without access to decent education, health care, and basic services. They work in the countryside as subsistence farmers and fishermen or in urban slums as construction workers and rickshaw drivers. The number, over 26 million new internally displaced people due to natural disasters, is difficult to grasp. It is above all individual fates, which make us sense the struggle for survival in which part of humanity already finds itself.

Migration Patterns in Times of Climate Change

Migration is woven into our lives like a fine tapestry. People move, find a new job, a new environment. Fewer and fewer people spend their retirement where they were born. But there are also very different types of migration, including expulsion, or resettlement—each with its own patterns, which can be broadly described along three continuums: the degree of voluntariness (was the exodus forced or did it occur by choice?), the temporal aspect (do people leave their homeland permanently, or only for a short time, and then return?), and the geographical dimension (what distance do people cover and do they cross national borders in the process?).

Between these different poles, there are mixed forms that largely preclude hard categorization and demarcation. Migration that was originally seasonal, for example, can gradually extend over an increasingly long period and ultimately become permanent. And the question of the extent to which migration actually occurs out of one's own free will, when poverty or a lack of prospects characterizes everyday life, often cannot be answered unequivocally. Most of the migration therefore tends to lie in the gray zone.

In a sense, climate migrants enter a risk exchange. The threat of hunger in the countryside due to crop losses may be exchanged for existential insecurity in the city, where there is no money to buy food. The risk of storm damage is pitted against the risk of having to live in densely populated slums, suffering from extreme heat under a corrugated roof. In fact, migrants are often worse off afterward than before. And yet, they take the risk, which is often supported by others: the family, the partner, the village community, which perhaps pool money to make the exodus possible in the first place. At the destination, there may be people who welcome the newcomer, a diaspora that helps with the arrival. Migration is thus not just the mere movement from A to B, but is part of a social system—it is the source of collective hope.

In addition to people who migrate on their own, there are also those who are resettled by government programs. Such measures usually lead to deep caesuras, destroy social ties, and cause structures in the communities to break down. It is not uncommon for those affected to even lose their economic livelihoods. Historical examples of extensive resettlement programs demonstrate how brutal this uprooting can be and the human rights violations that accompany it, especially when authoritarian regimes initiate it. In the German Democratic Republic (GDR), for example, resettlement of entire villages for coal mining in Lusatia left deep scars. Nevertheless, villages are still being demolished in Germany today in order to dredge coal.

The possibility to migrate is extremely unevenly distributed globally. With a German or American passport, at least tourist entry into many countries is possible without any problems. However, a large part of humanity has very limited freedom to travel and is equipped with passports that require complicated and often opaque visa procedures and rarely elicit a friendly "welcome" upon entry. Thus, migration routes are determined not only by the will and ability to migrate, but especially by borders that are politically determined and increasingly fiercely defended.

Decades of Climate Migration

Estimates of how many people will migrate in the future due to climate change vary widely, ranging from the assumption that climate migration is a myth to 1 billion displaced people by the middle of this century. Both scientifically supported and dubious projections are in circulation. Where does this uncertainty come from, and what figures and research are helpful in making policy decisions? In a pessimistic scenario, the World Bank projects over 200 million climate-induced internally displaced persons by 2050 across six world regions [10, 11]. Depending on how quickly emissions are reduced, the number could be cut in half or even further. While the World Bank report is among the most methodologically advanced of the cross-regional studies, many uncertainties remain. For example, it is highly questionable whether a shift of more than 200 million people can actually take place within national borders or whether this would unleash further migration dynamics across countries.

Simply extrapolating the present into the future is not possible because of the complexity of the movements. The further ahead one looks, the more difficult it becomes to predict climate migration dynamics. What development paths are taken now and in the coming decades? Will global warming slow down? If so, how much? Can poorer and rural regions share in global economic growth? What labor market regulations will come into effect? What educational and work opportunities will women have? By how much will the world population grow? These and other factors affect the dynamics of climate impacts and migration in fundamental ways. People will tend to migrate from low-lying areas to higher elevations to buffer increasing temperatures and protect themselves from rising sea levels.

Historically, people have always preferred to settle on coasts and along rivers and fertile deltas. It is precisely these areas that are now particularly at risk. The Netherlands shows how infrastructure measures, like dike construction, adaptive architecture, and designated floodplains, can keep areas habitable that would otherwise fall victim to floods. But not everywhere can such innovative measures be technically implemented or financially feasible. In poorer countries, where not even basic needs are met, it is difficult to imagine that investments in billions of dollars can be made in the next few years or even decades to protect areas. Especially around the tropical belt, large populations are defenselessly exposed to climate impacts due to poverty. In the long term, higher end warming scenarios could lead to a shift of the population from the tropical belt toward more temperate regions—with all the social upheavals that such a redistribution of humanity would entail.

Small Data, Big Data, No Data?

The fact that observations and projections on climate migration are characterized by uncertainties also has to do with poor data. Let us first look at the migration side. Around the world, migration is recorded at different points in time, primarily through national censuses, i.e., population censuses that usually ask questions about changes of residence. However, there is no uniform definition of migration. In some places, migration is considered to have taken place after 6 months of absence from the place of origin, while in others, it is only considered to have taken place after 12 months. Seasonal migration is often not registered at all. In addition, censuses are only conducted after long intervals because they involve considerable effort. In the United States, for example, only once every decade, a census takes place. The impact of short-term natural disasters between these time slices can therefore hardly be read from the data. Moreover, sometimes, the reasons for migration are not asked at all, and if they are, "natural disaster" is missing as an answer option. It is subsumed under "economic" or "other" criteria. In many regions of the world, censuses take place even less frequently and more irregularly. In remote areas, census coverage is poor at best. Yet, it is precisely there that climate impacts can play an important role in migration decisions. In addition to nationwide surveys, there are subnational surveys conducted at the local or regional level, for example. But here, too, there are many definitional issues that make comparisons across regions and over longer time periods difficult.

Migration researchers also collect data specifically for their studies through surveys or various interview formats, in order to answer specific research questions. A new data source is cell phones, from which metadata can be analyzed anonymously and thus provide extremely high-resolution information on migration after natural disasters. The first results generated in this way are already available [12]. This enables aid organizations and governments to obtain a better picture of the situation after a disaster. However, there is a bitter aftertaste: Especially in countries with weak data protection guidelines, this information is often analyzed by foreign teams of scientists. In the worst case, cell phone data could be misused by authoritarian governments to exclude, monitor, or persecute people—even if the anonymized data cannot be traced back to who exactly is using the device. In any case, such data should only be released under strict conditions.

A closer look at the data on climate impacts reveals major regional differences. For example, many developing countries have far less observational data than industrialized nations. In recent decades, weather stations have been dismantled in many places. This means that projections are less accurate and therefore less robust. In addition, climate research elsewhere is not as well equipped as in Europe and the United States. On the African continent, for example, there are only a few universities that offer climate physics as a subject and hence could train capable scientists. In addition, the national job market is not always promising for graduates of such courses [13]. Therefore, many countries depend on the knowledge produced in a few industrialized countries to develop their own climate adaptation planning. This uncertainty about regional climate impacts also affects climate migration research. But in both climate and migration, there is an increasingly dense database and a growing research community working to fill gaps and build solid knowledge about the complex interactions in and between the two fields.

> **Hot, Hotter, Africa**
> The Nigerian poet, trained architect, and Alternative Nobel laureate Nnimmo Bassey describes with shocking urgency how threatening the situation already is in his book *To Cook a Continent*. For him, the combination of the climate crisis and the exploitation of fossil fuels creates the perfect storm to destroy the ecosystems of the African continent and the people living there with it. For example, a 2 °C rise in global mean temperature could result in a much higher mean temperature rise regionally: 2.5 °C in North Africa and as much as 3 °C in countries like Egypt or Libya. A global increase of 4 °C would perhaps lead to over 7 °C for the region. It is hardly conceivable that the same number of people would still be able to hold out and survive there. A city like Cairo, for example, would face extreme difficulties to cope with such a dramatic challenge. The stakes are thus very high for the continent when it comes to the question of how much more we can contain climate change. We, that is, primarily people in the industrialized countries and the global middle and upper classes. In fact: The average CO_2 footprint in sub-Saharan Africa is 0.8 metric tons per year [14]—according to some calculations, it is lower than that of a large carnivorous dog in Europe. Many lifestyles in the region are climate neutral, in part simply due to poverty, but certainly also because of traditional lifestyles that are in harmony with nature. By way of comparison, the average CO_2 footprint in the United States is 14.4 metric tons and in the EU is 6.4 metric tons per capita and year. This equates to a colonization of the atmosphere, the pollution of the sky that belongs to everyone.
>
> In his home region, the oil-rich Niger Delta in West Africa, Bassey has witnessed how international corporations, in league with corrupt elites, exploit regions. Fertile soils are contaminated and rendered unusable in the long term

by leaking pipelines that are apparently not worth the effort to repair. Only a small number of people benefit from oil profits. Poverty is high despite the abundance of resources.

Whether we fill up our vehicle with gasoline, use heating oil to soften harsh winter cold, or buy petroleum-based plastics to meet our everyday needs, the truth is inescapable. We are all accomplices in this development—and inevitably so, because there is no escape from the system of oil-based economics unless we overcome it as a society. As individuals, we remain trapped.

The Nigerian population not only suffers from the direct consequences of oil production, but also from rampant heat, droughts, extreme precipitation, and rising sea levels. And as if that were not enough, more and more pressure is being exerted to repeat the West's mistakes in industrial agriculture—more pesticides and more fertilizers translate into less biodiversity and more land use emissions.

As an environmental activist, Bassey courageously opposes all this with his think tank Home of Mother Earth Foundation (HOMEF). He analyzes exploitative structures with sharp words and organizes local education programs on the subject of sustainability. Growing civil society involvement of this kind can create prospects for a different, self-determined development of the continent.

The Nomadization of Our Way of Life

Although it is not yet clear exactly how severe future climate impacts will be, there is no doubt that shifts in the climate system will make migration an inevitable means of adaptation. People will work seasonally in agriculture in more fertile areas or leave particularly hot places during the summer months. Many residents of areas particularly affected by climate change will be forced to migrate by the complex interaction of poverty, demographic trends, and climate impacts. Many people are already permanently on the run. Displaced from their homes, they are relocating to other areas where they live in makeshift shelters or in huge refugee camps, such as Dadaab in Kenya. Returning home is becoming increasingly difficult, because their livelihoods have been destroyed by conflict and because climatic stressors make it even more difficult [15].

Some migrants I spoke with could not say how long they would stay in one place. They planned to hold out until a metaphorical or actual physical storm snatched away their livelihoods again and then move on to the next town or village. The constant change and the great uncertainty have become entrenched in their lives.

Increasing nomadization is a global process. This trend is also illustrated by cosmopolitan Parag Khanna in his book *Move*. In it, he describes with fateful optimism the growing willingness, but also the necessity, to migrate [16]. Today, more flexibility is provided and required in a number of occupational sectors. Less stable employment and volatile world markets have introduced greater uncertainty into urban lifestyles. As a result, the relocation of people is not always voluntary. Climate change remains the great uncertainty in human mobility projections. It could soon become a game changer in global migration movements and even eclipse previous upheavals.

What Climate Migration Has to Do with Us

Even if other countries are harder hit by climate change than the United States or Europe, people in industrialized countries can hardly isolate themselves from dangers that unfold in seemingly faraway places. The COVID-19 pandemic has shown this in all clarity. The overexploitation of resources, such as that caused by the wildlife trade in China, ultimately also affects people in Pittsburgh, Munich, or rural Colorado. Conversely, lignite mining in Lusatia and North Dakota affects livelihoods in the Mekong or Ganges-Brahmaputra deltas. Our lifestyle casts long shadows—as far away as the Marshall Islands in the Western Pacific and Bangladesh in South Asia. People most affected by climate change have contributed little to its creation. Recognizing these interdependencies is the only chance for human civilization. In this respect, the narratives in the upcoming chapters from different countries around the world also contain warning signals and stop signs. The risk of losing of an entire culture, for example, in some flat-lying island states, should not only concern us, but move us to a rapid rethinking and reorientation. A "business as usual" leads directly against a mighty wall—and no one wants to experience this impact, regardless of political convictions or level of prosperity.

If all 25-year-olds were to inherit the CO_2 debts of their forefathers and mothers since the beginning of industrialization, this would result in extremely high per capita emissions in the industrialized countries. Germany has a high historical emissions burden, as do other countries such as the United States, Belgium, Great Britain, and the Czech Republic. However, it is particularly alarming that China, with its large population and relatively short fossil fuel industrial history, has also rapidly caught up with historical per capita emissions. If all countries followed, for example, Germany's development, the Earth system would be transformed into a completely different state. The climate crisis would become a climate catastrophe. Already today, the environmental changes are serious, especially in those countries that have contributed little to global emissions, i.e., where the historical CO_2 emissions are low.

How internal migration affects international migration as a result of these changes is not always immediately apparent. Even if the migration stories of people arriving in Europe or the United States hardly tell of superstorms or floods, their migration may be indirectly related to increasing climate impacts via cascading effects. For example, migration from rural areas to nearby medium-sized cities may lead to a competitive situation in which newcomers undercut local workers out of necessity with very low wage demands. They may then migrate to the capital, where this situation is repeated. As a result, well-educated skilled workers there feel pressured by the migrants and emigrate abroad. Nevertheless, the trained engineer will hardly see their migration in connection with climate impacts, even if there are such indirect chains of effects. It is indisputable that, due to demographic change, prosperity in many industrialized countries cannot be maintained in the long term without immigration.

But whether the cascading effects just described occur at all also depends on labor market dynamics. If there is a great need for labor, companies will likely vie to keep bright minds. However, many countries, especially in sub-Saharan Africa, currently have high youth unemployment. Young people who could use their full potential for a better future are forced to keep their heads above water with odd jobs on the black market. The chances of economic advancement for low-wage workers are very poor in many cities around the world. Informal settlements are just one example of urban poverty spirals in which migrants are often the last link in the chain.

The effects of climate change on our social system have not yet been sufficiently researched. But there are indications that extreme heat increases the propensity to violence and crime rates [17]. Such secondary effects, which cannot be directly attributed to climate change, are becoming further drivers of migration.

Caught Up in Climate Chaos

Not being able to move out of harms' way when a threat occurs can be even more disastrous than being forced to migrate. Climate change may cause increased immobility among certain populations. For example, when people lose everything in a storm, they sometimes do not have the resources to flee. Vulnerable groups in particular, such as the elderly, the extremely poor, people with disabilities, or those who have sustained injuries, are often affected by involuntary immobility. They are trapped in climate chaos, in places where there is nothing left but life itself.

The terrible incident in 2021 during the floods in the German Eifel region, when 12 people died in a facility for people with disabilities because they were not brought to safety quickly enough from the flood wave at night, is just one example of how much the degree of risk depends on factors such as place of residence, gender, physical abilities, ethnicity, and nationality. These factors can make the difference between life and death.

At the same time, there are people who do not want to leave their homeland even in the face of extraordinary risks. The loss of their traditions, the loss of the land on which they grew up and in whose soil their ancestors are buried, weighs more heavily for them than the risk of dying in a devastating storm. They insist on their right to stay, come what may. Some also suspect that their lives could get worse if they migrate. This is what my colleague Himani Upadhyay told me about after returning from a field research trip to the northern Indian state of Uttarakhand at the foothills of the Himalayas. The women in the remote areas made a conscious decision not to migrate, even though the work was becoming increasingly arduous and glacier melt as well as changing weather patterns were affecting their traditional agriculture [18]. Trading a difficult life back home for mere survival in a slum was not an option for them. The desperate struggle of people against the forces of nature to defend their own homeland is also part of this book.

Corona and the Climate

Meanwhile, in India, where my journey on climate migration began, something previously unimaginable happened during the coronavirus pandemic. Migration flows on the subcontinent were reversed, at least briefly. Millions of migrant workers lost their jobs and income due to lockdowns and infection control measures and returned to their home villages. There was, despite great poverty, at least the possibility of self-sufficiency through the cultivation of grain and other food crops. At the same time, the mass return migration contributed to the spread of COVID-19—a fatal consequence because health-care access in remote areas is usually even worse than in the cities. Thus, coronavirus hotspots emerged throughout the country. The temporary reversal of migration flows in India illustrates that external shocks have unpredictable and widespread consequences. While most scientists currently believe that climate change will amplify existing migration dynamics such as rural-urban migration, entirely new corridors may also emerge when unprecedented extreme events occur.

The question is what will people do who can still afford a piece of normality? Do they try to maintain the status quo or do they prepare themselves for a future determined by climate change and confront the challenges that come with it? Are we moving or standing still in the floodlights of the climate crisis?

Future Prospects

What might global climate migration look like in the future? Recognizing that many other developments are also conceivable, I would like to present two possible scenarios for the second half of the twenty-first century to illustrate a pessimistic and an optimistic outlook. These two scenarios serve to demonstrate the dimension of the problem and the possible range of developments.

Scenario 1. The Fall Into the Abyss
Tipping cascades, population growth, and tightened border regimes.

If several of the Earth's key ecosystems were to collapse, the livelihoods of countless people would be threatened. Food, health, security, shelter—all these would be at stake. Subsistence farmers in large parts of the world would not be able to grow their crops as before. Many areas would become uninhabitable due to persistent heat extremes, frequent tropical cyclones, and rising sea levels. Coastlines and islands could no longer be adequately protected because the multitude of disaster hotspots would make countermeasures largely impossible, not to mention the material, financial, and political resources that would quickly be depleted.

The scenario continues to worsen. Serious conflicts over access to basic services would be unavoidable because the population has grown considerably in recent decades. If resources are available at all, they would have to be distributed among more and more heads. Dwindling livelihoods in rural areas would subsequently lead to increased rural-urban migration. But if the urban labor market lacked capacity, those who did migrate would be caught in a spiral of poverty. Slums would continue to grow into areas already vulnerable to climate change, such as mountainsides or flood zones. Thus, misery would gradually spread to inner cities, which would then become increasingly violent. Some countries would therefore also try to stop internal migration to suppress the growth of informal and hard-to-govern areas. Many people would be forced to remain in danger zones or attempt to cross their country's borders. However, many states could tighten border regimes, make entry more difficult, or even prevent it altogether, so that more and more people would lose their lives.

In such a negative scenario, what has already happened to some healthcare systems during the COVID-19 pandemic would occur on a larger scale: If prevention fails—or was never seriously pursued—and certain critical thresholds are crossed, damage can only be contained, but no longer averted by adaptation measures taken too late. Exponentially growing crises then overwhelm even industrialized countries. Migration as a survival strategy would reach unimagined proportions, changing its dynamics, and could end in chaos and violence.

> **Scenario 2. The Dance on the Cliff**
> *Paris climate agreement is mostly realized; the world population stabilizes; climate impacts are recognized as a reason for asylum.*
>
> If the Paris Climate Agreement is adhered to, so that temperatures could stabilize at 1.5 to a maximum of 2 °C above the pre-industrial level, far-reaching options to cope with climate impacts would remain open. For example, sustainability-oriented modernization of agriculture could safeguard livelihoods in rural areas, even if rainfall patterns change. Reforestation measures would buffer warming locally and could also improve water availability. Severe climate impacts would remain limited to particularly fragile ecosystems, such as tropical coral reefs. Island states and coastal zones could be largely protected through major infrastructural adaptation. Nevertheless, some resettlements would still be necessary, for example, away from very low-lying areas threatened by sea level rise. Some people would have to be provided with separate protection rights, enabling them to live and work in other countries without a visa after losing their homeland.
>
> Investments in renewable energy could promote polycentric urbanization through decentralized energy production, which in turn would ensure a more even distribution of immigrant populations across multiple destinations and generate new labor markets. At the same time, energy exports could be reinvested into education and development. If this resulted in less accelerated global population growth, population pressures in densely populated areas would also be a less relevant factor in migration decisions. Thus, while climate migration would remain an issue for certain regions, its negative impacts could be mitigated through development policies, making it an effective form of adaptation to climate change for some populations.

Thesis

The climate crisis has become the new normal, even if its extent can still be limited. Migration enables those affected to survive, but not necessarily to maintain their standard of living.

Chapter 2
The Right to Stay and the Freedom to Go: Legal Aspects of Climate Migration

The Refugee Convention and Its Limitations ♦ The Principle of Non-refoulement ♦ Climate Displacement Within Countries ♦ Holes in the Protections Afforded by the Universal Declaration of Human Rights ♦ Climate Lawsuits and the Human Right to a Clean Environment ♦ Ecocide: A Crime Against Peace? ♦ State Without Territory?

During my field research in Bangladesh, people recounted with horror how bullets followed them as they ran in fear of death while trying to cross the border with India. They describe the silent search for refuge from patrolling guards. Tropical cyclones had destroyed their homes. Every year, dozens of Bangladeshis die on the Indian border, many of them migrants who risk everything for a better life. They try to cross the border illegally because they have virtually no access to formal visas. With a European or US passport entering India or Bangladesh, it usually only requires filling out a series of forms and paying a processing fee. For those living at the poverty line in Bangladesh, however, borders present hazardous and often insurmountable hurdles. The dividing line between India and Bangladesh is only one of the many deadly border crossings on our globe.

Borders are hotspots for human rights violations. Their ruthless defense has become a political point of contention in our society. "Even democracies are capable of using brute force to keep refugees from entering their borders," states political scientist Gerald Knaus [19]. As a historical example, he cites Switzerland's rejection of tens of thousands of Jewish refugees during World War II. Even though many records were destroyed, it can be assumed that a large part of these European migrants were subsequently murdered by the Nazis. Numerous historical and contemporary examples show that closed borders violate human dignity and, in the worst cases, lead to loss of life and of human potential.

Australia is interning refugees in Papua New Guinea, and Europe is expanding its border protection by enlisting other states in limiting migration from Northern Africa and the Sahel. But what consequences do the existing border regimes and asylum laws have for climate migrants? For people who must migrate due to climate impacts, international borders are considered especially hard to cross. This is because these migrants lack both a specific, climate-related protection status and a right to asylum. This is also linked to the definition of the term "refugee," which was included more than 70 years ago in the Geneva Refugee Convention, the "Convention relating to the Status of Refugees."

The Refugee Convention and its Limitations

The Geneva Convention is the cornerstone for the protection of refugees. Created in response to large-scale transnational displacement during World War II and the persecution of Jewish people and other groups by Nazi Germany, the convention established persecution on political, racial, social, or religious grounds as the criterion for entitlement to protection. Accordingly, a refugee is a person who:

> owing to well-founded fear of being persecuted for reasons of race, religion, nationality, membership of a particular social group or political opinion, is outside the country of his nationality and is unable or, owing to such fear, is unwilling to avail himself of the protection of that country; or who, not having a nationality and being outside the country of his former habitual residence as a result of such events, is unable or, owing to such fear, is unwilling to return to it [20].

As can be seen from the definition, the criteria for obtaining refugee status do not apply to people seeking protection from the impacts of climate change because they are not politically or otherwise persecuted. However, if climate impacts are so severe that they destabilize a country and undermine the rule of law, for example, it is conceivable that the Geneva Convention could apply. In this case, however, it would not be the physical consequences of climate change as such that would be recognized as a reason for flight, but the resulting secondary harm to the political or social system, which could lead to the persecution of individuals [21]. However, the relevant standard of proof is extremely high. Not even war refugees automatically qualify under the Geneva Convention. Thus, there would have to be a risk of genocide, for example, due to secondary damage resulting from climate impacts, or certain groups would have to be systematically persecuted in order to qualify as "refugees" under the Geneva Convention. Since the concept of refugee is so narrowly defined and, at least theoretically, comes with separate protection rights, the majority of scientists do not speak of climate refugees, but of climate migrants, or describe the movement with the bulky, but somewhat more pleasing term of "human mobility in the context of climate change."

Even though the characteristics of many types of migration often equate to flight-like movements, the reasons for moving play a decisive role in the right to protection. Thus, persecution by one's own state is a legally recognized reason for obtaining protection in another country, but life-threatening poverty or an all-destroying storm is not. And another point must be considered: While the vast majority of today's climate migration is limited to movements within national borders, it can by no means be ruled out that as climate damage increases, attempts to cross borders will also increase. In view of the distribution of the sources of climate harm, i.e., global greenhouse gas emissions, it seems legitimate for people whose living environment has become uninhabitable due to climate impacts to seek protection in countries that are largely responsible for them.

An ethical responsibility to protect is derived here not only from the humanitarian imperative, i.e., the requirement to help people who are in distress; it also arises from the polluter-pays principle. In essence, this means that those who cause harm must pay for it. Forcing people to migrate may be a consequential harm of global warming. Enabling dignified migration as an adaptation to climate change is thus a matter of climate justice. But this moral obligation is not yet implemented through a legal mechanism that would entitle victims to such migration. In this regard, applicable law and justice are currently far apart.

There is a good reason why changes to the Geneva Convention are not demanded even by the most determined activists. The Geneva Convention in its current form is under constant attack by politicians. Therefore, there is a great danger that, in the course of such a renegotiation, further restrictions on existing refugee protections would be enacted rather than extensions of protection to other groups, such as climate migrants. The closing of the EU's external borders by the use of force is a bitter reality.

The Principle of Non-refoulement

One of the central rights under the Geneva Refugee Convention is the right to non-refoulement:

> No Contracting State shall expel or return ("refouler") a refugee in any manner whatsoever to the frontiers of territories where his life or freedom would be threatened on account of his race, religion, nationality, membership of a particular social group or political opinion [22].

In addition, according to the Universal Declaration of Human Rights:

> Everyone has the right to seek and to enjoy in other countries asylum from persecution.

And international law stipulates that people may not be deported or returned back to their home country if they expect human rights violations there. Nevertheless, affected persons and NGOs repeatedly report "pushbacks," i.e., the forcible pushing back of migrants at the external borders of the EU. From the pushbacks at the Greek and Croatian borders, for example, there is evidence of sexual violence and the use of whips and batons. Respect for human dignity and applicable law are thus undermined by a European border policy that condones violence and dehumanization of those seeking protection. At the very least, Fabrice Leggeri was pressured to resign as Frontex chief because he wanted to cover up the unlawful practices. He vacated his post in April 2022 [23].

In 2021, the European Court of Human Rights ruled that the refoulement of an Afghan family at the Croatian external border 4 years earlier was unlawful. The case had attracted particular attention because a 6-year-old girl was hit by a train after border police ordered the family to walk back along the tracks at night toward Serbia [24]. Nevertheless, pushbacks continue, for example, in 2021 on the Polish-Belarusian border. Thus, the Belarusian Lukashenko regime had created incentives for refugees from Syria and Iraq to travel to Minsk, from where they would be further piloted to the EU. The migrants were practically used for a new form of warfare, with the aim of putting pressure on the EU, which had previously imposed sanctions on Belarus. The EU, however, did not go along with this.

The Polish government—with the backing of the EU—showed its toughest side and rejected even those entitled to protection. The top priority was not to succumb to the power games of the Belarusian autocrat. However, it was the refugees who suffered as a result of this confrontation: They had to endure the bitter cold in the border area, and their applications for protection were turned down.

When guilty verdicts are reached on charges against such inhumane acts, usually only individual border police officers are suspended or dismissed. The political strategy behind the pushbacks remains unaffected. But the pillar of refugee protection falters when migrants are politically instrumentalized. The cruel logic of demonstrative rejection for the purpose of deterrence runs counter to the values that are proclaimed as a high good within the borders of the EU.

What role does the principle of non-refoulement play in the case of severe climate impacts? Some lawyers argue that the principle of non-refoulement can also be triggered by climatic risks, for example, when there is a risk to life. The UN Human Rights Committee generally recognizes this possibility, but so far, attempts to judicially invoke the principle in this context have failed [25]. Proving that climate impacts threaten an individual is difficult because it is very costly to establish legally watertight causal chains between damage to the environment and a direct threat to life posed by a return home. In addition, the burden of proof currently lies with those affected. To cite generally difficult living conditions due to climatic changes, which also affect the general public, is not sufficient as a reason to be granted asylum [26]. Rather, the concrete threat to the individual must be proven. Illegal pushbacks can also prevent migrants from even getting the chance to have their asylum applications examined in the country of arrival. Moreover, the possible creation of a right to climate asylum does not necessarily guarantee its enforcement.

The existing gaps in the protection mechanism mean that those people who try to cross national borders to find a new place to live due to natural disasters have no legal right to admission. If they do manage to cross borders, their legal situation in the countries of arrival often remains unresolved for years. Some states even use internment camps to "detain" people whose residency status is uncertain. The quality of life in such mass shelters is often lower than in traditional prisons. Some countries move the shelters abroad to create an even stronger deterrent effect. For example, Australia operates detention centers for refugees in Papua New Guinea and on the small island nation of Nauru. Asylum seekers from Afghanistan, Iran, Bangladesh, or Pakistan are held there for years. During their imprisonment, they are not allowed to move freely and experience humiliation or abuse [27]. Many of those housed there take their own lives. Their alleged crime: migration—the attempt to reach safety, to flee persecution, hunger, or oppression.

Climate Displacement within Countries

Since migration in the context of climate change and global warming largely takes place within national borders, the protection of internal migrants is particularly relevant because chaotic situations often arise when many people must leave simultaneously to ensure their survival after a devastating storm, for example. This was the case in the Caribbean in 2017, when tropical storms unleashed their destructive force on several islands, or in the Philippines when Typhoon Odette displaced hundreds of thousands of people in December 2021. For months after the extreme events, vast areas were still underserved, and those who remained faced existential challenges. Such disasters often form an enduring deterioration in the lives of those affected. Due to the severity of the damage, many must settle in a new location for an indefinite time period.

Certain groups need special protection. In particular, children who have lost their parents or women in distress become targets of organized crime and human trafficking. When governments are unable to help their populations in a situation of collective shock, a power vacuum is created that is often exploited by third parties. To address such threats and commit states to a minimum level of protection, Guiding Principles for addressing internal displacement were developed at the international level in 1998. They include groups of people forced to flee due to (man-made) natural disasters. In contrast, slow-onset climatic changes, which can also lead to forced displacement, are not explicitly mentioned.

In addition, the Guiding Principles are not binding, but are formulated and implemented voluntarily by individual states in legal documents. Accordingly, they are poorly implemented in countries where there is large-scale internal migration and forced displacement. They are not sufficient as an instrument to protect climate migrants. However, despite the lack of a uniform solution to the plight of internally displaced persons, the document generated change in various regional frameworks that reference the Guidelines. This is true, for example, of the African Union's Kampala Convention, which focuses on the protection of internally displaced persons on the continent. Regional agreements, in particular, can create more security for displaced persons in the absence of consensus across the international community and, in turn, provide an impetus for new international legal agreements. Thus, progress in soft law, i.e., nonmandatory law, also affects national "hard" legislation and drives innovation.

If there is no progress in binding international law, nonbinding, regional agreements can gradually pave the way for change. The 1984 Cartagena Declaration, for example, agreed to by Latin American states, defines the concept of refugee more broadly than the Geneva Convention and includes as legitimate grounds for flight situations in which public order is seriously compromised. It was first applied by Brazil in the Venezuelan refugee crisis. It could also become relevant in the context of climate change.

At the global level, two milestone documents were debated over the years and finally adopted in 2018: the Global Compact for Safe, Orderly and Regular Migration [28] and the Global Compact on Refugees [29]. The latter includes only a brief reference to climate change itself not triggering refugee movements, but "interacting with drivers of refugee movements." This corresponds to the narrow definition of refugee that also underlies the UN Refugee Pact.

The UN Migration Pact, on the other hand, addresses climate change and environmental change in more detail and points to measures that can be taken to reduce the need for climate migration, such as better disaster preparedness and access to reliable information. This pact is also nonbinding and only sets broad goals for improving the situation of migrants in countries of arrival and mitigating negative drivers of migration. Nevertheless, the United States, Israel, Hungary, Poland, and the Czech Republic voted against it in the UN General Assembly. Migration, especially when it means movements from developing to developed countries, is often politically instrumentalized and used for polarization. This is one reason why progress in rights development has been minuscule, while the challenges continue to grow. Yet, organizations such as the German-supported Platform on Disaster Displacement, a states' initiative, continue to try to achieve through international dialogue better protection on a voluntary basis for persons displaced by cross-border disasters. The Platform's predecessor organization, for example, developed a detailed agenda that identified areas where improvements could be made [30]. The UN Migration Pact also makes explicit reference to this agenda.

Holes in the Protections Afforded by the Universal Declaration of Human Rights

All migrants, regardless of their place of residence, are entitled to the protection of their human rights. Migration is a human right and is enshrined in Article 13.2 of the Universal Declaration of Human Rights:

> Everyone has the right to leave any country, including his own, and to return to his country.

Nevertheless, the reality shows, on the one hand, that migrants in particular are exposed to human rights violations and, on the other hand, that many of those affected do not even have the possibility to claim their rights or even to go to court. Apart from that, especially poor population groups, who have or had hardly any access to education, often do not know about their rights.

In migration studies, we speak of the cycle of migration, which begins with preparations and departure, then concerns migration itself, and finally culminates in the arrival and integration of people. At each stage, the rights of the individual should be protected [31]. However, this is not easy to ensure, because during these different stages, individuals are exposed to various risks, and there may also be several states and institutions responsible for protecting their rights. The rights of indigenous people also stand out in this context, as many of those on the front lines of climate chaos belong to indigenous groups. For example, even in the Arctic Circle, the environment is changing rapidly, threatening the culture and way of life of the entire populations [32]. The diversity of migration contexts requires legal protection instruments that do justice to this diversity [33].

When powerful industrialized countries flout applicable law, as in the case of the pushbacks at the EU's external borders or in Australia's detention camps, there is often little recourse to prevent this. Only through public pressure or targeted legal proceedings for demonstrable human rights violations can changes be brought about. But the latter, in particular, can drag on for years, long enough for lives to be derailed or destroyed.

Climate Lawsuits and the Human Right to a Clean Environment

Despite the ongoing stagnation in legal developments in favor of the protection of climate migrants, there have also been positive changes in the field of climate and law in recent years. A whole series of initiatives and landmark rulings are steering climate protection in the right direction (Fig. 2.1).

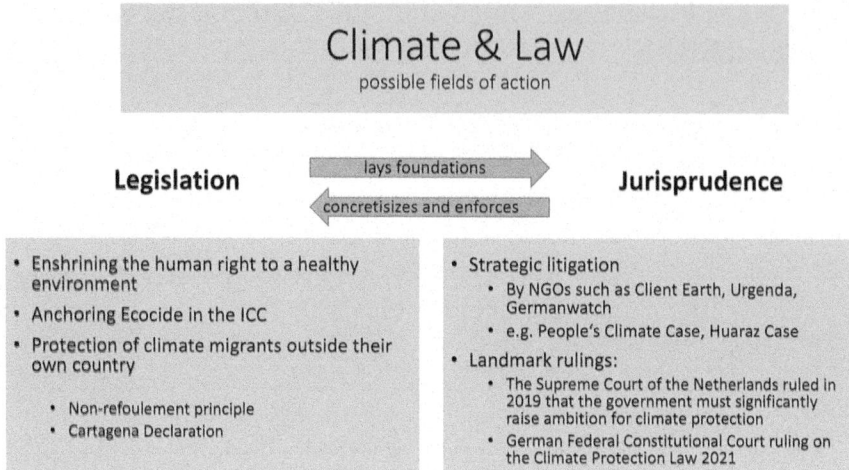

Fig. 2.1 Possible fields of action in the area of climate and law, own illustration

In October 2021, for example, negotiators who have been advocating for the right to a clean environment at the UN level for years achieved a first breakthrough. The UN Human Rights Council, based in Geneva, recognized the right to a "safe, clean, healthy and sustainable environment" as a human right [34].

This decision is regarded as an important signal for the further development of national law, even if the decision does not yet create a legal obligation. Multinational agreements often only develop their effectiveness over the years, but they set new standards for legal evolution at various levels. For the area of climate and migration, the decision could have the effect of giving more importance to the prevention of environmental damage, since there is now a direct link between the claim to an intact environment and human rights.

Another positive sign is that, in several countries, lawyers have formed groups that want to achieve more climate protection by means of strategic negotiation. One example is the nongovernmental organization Urgenda, which sued the Dutch government on the basis of the precautionary principle—and was successful. The case went all the way to the highest court in the Netherlands, and in the end, the government had to increase its emission reduction targets and commit itself to taking preventive measures to avert harm to its citizens. This includes more ambitious climate protection. By legally forcing states to do more to protect the climate, lawyers may save us from becoming climate refugees in our own country. It is not only the individual case that counts, but also the overall dynamic that can unfold through media attention and standard setting.

The German Federal Constitutional Court also issued a landmark ruling in 2021, obliging the German government to tighten up its climate protection law. The reason given for the decision was that the planned implementation of climate protection measures to comply with the 1.5 °C limit would come at the expense of younger people. If more ambitious mitigation measures were not taken early, massive emissions reductions would have to be implemented starting in the 2030s. "The regulations irreversibly postpone high emission reduction burdens to periods after 2030," [35] reads the Constitutional Court's press release. It clarifies that postponing significant emission reduction measures until after 2030 results in disproportionate restrictions on the liberty of younger people. The liberty rights of younger people would be curtailed because virtually all areas of life are affected by CO_2 emissions. At the same time, the judges recognize that the longer emission reductions are delayed, the more far-reaching the encroachment on civil liberties will be in the future.

With its ruling, the Federal Constitutional Court not only set a benchmark for greater intergenerational justice, the judges in Karlsruhe also defined the temperature limits enshrined in the Paris Agreement as a guideline for political action in Germany. If it becomes apparent that the climate consequences associated with the aforementioned temperature limits will be more severe than expected, the legislature may have to take even more stringent measures to mitigate risks.

It is precisely the lower temperature limit of 1.5 °C, which is mentioned several times in the ruling, that is essential with regard to migration and displacement. This is because the higher temperatures rise on Earth, the more people will be forced to leave their homes. A warming of more than 1.5 °C would make low-lying island states and some coastal areas uninhabitable. The factual report of the ruling also refers to this connection: Germany could be "indirectly affected by the consequences of climate change in other parts of the world through the increase in climate-related flight and migration to Europe." [36].

Decisions by independent courts are generally playing an increasingly important role in climate protection. The lawsuit filed by a Peruvian farmer against the German energy giant RWE could become an important precedent—if the judges rule in his favor. Saúl Luciano Lliuya lives in Huaraz, a town in the Peruvian Andes. As the glaciers of the Andes melt, the villages on the mountain slopes are exposed to considerable risks. If an avalanche crashes into the glacial lake, violent flash floods can occur. Just how dramatic such glacial lake bursts can be is shown in images that went around the world from Uttarakhand in India in 2020. The flood wave at the foothills of the Himalayas swept 200 people to their deaths.

A similar scenario is also conceivable in the Peruvian Andes. The survival of the people and their stay in the traditional areas is no longer secured by the climate crisis. That is why the farmer from Huaraz sued the energy company RWE with the help of the nongovernmental organization Germanwatch. Lliuya's demand: The company should contribute to the costs of adaptation measures in proportion to its share of global greenhouse gas emissions. One adaptation measure, for example, would be the construction of dikes and protective walls that would prevent human lives from being endangered by the melting glaciers or residents from having to leave their villages to seek safety.

RWE is responsible for 0.5% of *all* man-made greenhouse gas emissions worldwide. In the event of a ruling in favor of Luciano Lliuya, a sum of only 21,000 euros would be due—little money for the energy company. But if the plaintiff were to be proven right, there would be a wave of lawsuits of unforeseeable proportions against major emitters such as energy companies. The Essen Regional Court, which had jurisdiction in the first instance, dismissed the lawsuit, but the Peruvian and his lawyer Dr. Roda Verheyen appealed to the Higher Regional Court in Hamm [37]. As part of the hearing of evidence, the court first scheduled a site inspection in Peru to examine the extent to which the plaintiff was affected by the risks of flash flooding. The judges and experts traveled to Huaraz in May 2022, and the outcome of the trial is still open [38].

Climate lawsuits, however, are often dismissed, such as the "People's Climate Case." Ten families from Germany, Portugal, France, Italy, Romania, Kenya, Fiji, and Sáminuorra, an association of young Sami people from Sweden, supported by lawyers and nongovernmental organizations, took the case to the European Court of Justice in order to force the EU to reduce emissions more drastically. The EU's target at the time of achieving a 40% reduction from 1990 levels by 2030 was too low to halt global warming under the Paris Agreement, the lawsuit said. Among the plaintiffs was a family based on Langeoog who worried whether their children would still be able to live and work on the North Sea island in the future. Higher storm surges and the possible salinization of the scarce drinking water resources acutely threaten the residents. All the families and the Sami Youth Association have been affected in different ways by the consequences of climate change, whether economically, culturally, or in terms of health. However, their lawsuit was already rejected on procedural grounds, and the plaintiffs were denied standing. According to previous case law, plaintiffs must be uniquely affected by EU legislation, in this case the too weak emission reduction laws. After all, all people are affected by climate change: "The applicants have not established that the contested provisions of the legislative package infringed their fundamental rights and distinguished them individually from all other natural or legal persons concerned by those provisions [...]. It is true that every individual is likely to be affected one way or another by climate change, that issue being recognized by the European Union and the Member States who have, as a result, committed to reducing emissions. However, the fact that the effects of climate change may be different for one person than they are for another does not mean that, for that reason, there exists standing to bring an action against a measure of general application [...]." [39].

From a purely moral point of view, this reasoning is a paradox, because the more serious the consequences of the failure to protect the climate, the more people are affected, but they are not protected by the judiciary either [40]. Nevertheless, there is one bright spot in the ruling. Through massive public pressure, to which the People's Climate Case also contributed, the EU's emissions reduction target was raised to 55%.

Ecocide: A Crime against Peace?

Damage to large ecosystems can have devastating consequences for the population in individual regions of the world and, via the finely woven Earth system, for humanity as a whole. The destruction of the Amazon rainforest to create land for soy production and cattle breeding through systematic slash-and-burn clearing illustrates the extent of the threat to the planet. When economic interests are in the foreground, environmental protection is often bypassed by companies and governments or given only scant consideration. Environmental catastrophes can have consequences that last for decades. This was the case with the explosion of the Deepwater Horizon oil platform in 2010, which released almost 800 million liters of oil into the Gulf of Mexico and resulted in record compensation payments [41]. Furthermore, 20.8 billion went to the affected US states and to the National Ecosystem Restoration and Research Programs. Although the amount of damages awarded was extraordinarily high, it is questionable whether this actually compensated for the losses. What is the cost of a fish's life or an intact ecosystem? Does it have any monetary value at all? One thing is certain: To this day, the northern Gulf of Mexico has not fully regenerated [42].

Even before the amount of the damages and fines had been quantified, the then CEO of the BP Group, Tony Hayward, announced to major shareholders that dividends would also be paid out in 2010—which was ultimately averted only amid great political protests. But the career of Hayward, who received a huge severance package from BP, was not harmed, nor was the fact that he faced massive criticism because BP had taken insufficient measures to prevent and repair oil leaks. Although he was forced to resign as CEO of BP, he quickly moved to coal giant and commodities company Glencore, one of the world's largest CO_2 emitters [43]. The company has been embroiled in various corruption scandals [44].

Should senior company officials who may be responsible for dramatic environmental disasters due to a lack of preventive measures be liable to personal prosecution? When is the negligence causing a disaster egregious enough to constitute a criminal wrong, and when is it simply an accident? Here, opinions vary widely.

Many environmental crimes are not even punished, even if long-term damage remains and people are displaced by them. Immediately, politicians whose irresponsible actions destroy pristine nature come to mind: former Brazilian President Jair Bolsonaro, for example, who was fueling the deforestation of the Amazon rainforest for agricultural profits. However, the justified finger-pointing at decision-makers should not ignore the industrialized countries, which contribute to the ecocide to a considerable extent through the high demand for cheap soy products as animal feed.

Ecocide includes serious and wanton destruction of ecosystems, which could also be punished by an international court or alternatively by the courts of individual states. A growing number of lawyers are campaigning for this. Their goal is to ensure that ecocide is punishable by the International Criminal Court (ICC) based in The Hague. Its mandate is regulated by the Rome Statute and covers the most serious crimes committed by individuals in international criminal law: genocide, war crimes, crimes against humanity, and crimes of aggression, i.e., violations of the prohibition of aggressive war. The International Criminal Court becomes active only when the crimes are not sanctioned by national courts. But not all states in the world are members and subject to its jurisdiction. Large states such as China and India have not joined at all. The United States and Russia signed the Rome Statute but ultimately did not ratify it.

A number of groups advocate for an expansion of the mandate and membership of the ICC. A group of lawyers led by Polly Higgins, the Scottish "Lawyer of the Earth," who died in 2019, have been engaged in a long struggle for the integration of ecocide into the Rome Statute. With the help of Stop Ecocide International, a nongovernmental organization founded by Higgins, an international panel of experts was convened to develop a definition of ecocide under international law. The panel submitted its joint draft in 2021. It defines ecocide as "unlawful or wanton acts committed with knowledge that there is a substantial likelihood of severe and either widespread or long-term damage to the environment being caused by those acts." The world-renowned jurists also made concrete suggestions on how to expand the International Criminal Court's mandate. Pope Francis has also spoken out on the subject of ecocide, indirectly advocating the integration of the crime into the Rome Statute [45].

Ecocide could have been included as a criminal offense much earlier. In the discussion drafts surrounding the creation of the International Criminal Court, serious environmental damage was proposed as one of the crimes against humanity as early as 1986 [46]. Until the mid-1990s, i.e., until shortly before the decision on the statute, this additional criminal offense was debated, but three European states ultimately prevented the inclusion of ecocide: the Netherlands, Great Britain, and France. For a long time, further debate on this important issue seemed futile, but in recent years, a new momentum has emerged.

State Without Territory?

Should it come to the horror scenario that the entire atoll states lose their territory due to sea level rise, not only humanitarian and ethical questions arise, but also legal ones: Will the state in question still exist? Do the inhabitants of the islands become stateless persons? What about the law of the sea? Will the sovereignty of the state over the exclusive economic zone (EEZ) disappear? Even if it is still possible to limit global warming, the threat to the low-lying island states has become so real that legal scholars and political scientists have already addressed these questions.

In general, statehood presupposes a territory, a population, and state power with a minimum of autonomy [47]. All three core elements of a state can be compromised by climate change: disappearance of territory through sea level rise, depopulation through transnational climate migration, and undermining of state power through extreme and recurrent natural disasters and the resulting social as well as political divisions. These changes in statehood are interrelated and may be mutually interdependent.

While the undermining of state power can take place via many widely ramified chains of effects and could only be attributed in part to climate change, the migration of a state's population is clearly attributable to it. Even before the overall loss of land area, people would leave their country [48]. In some atoll states, this dynamic may already have begun. The outcome, whether or not all people will have to migrate, is uncertain and is in the hands of the industrialized nations. If more and more people try to migrate, this could also weaken the ability of the state as a whole to defend its existence [49].

Contrast this with examples of powerful diaspora populations that both influence political events in their country of origin and are a strong voice for their compatriots at their destination. For example, Cuba's exiled population in the United States influences US foreign policy regarding the island nation. It is conceivable that a strong diaspora could demand rights for their home states. This is already the case in some places. Many climate activists from Pacific island states now live in Australia, New Zealand, or the United States for various reasons and exert pressure on the international community from there.

According to the current legal situation, if a state no longer exists, there can no longer be any citizenship of that state [50]. If the persons concerned were then not granted another nationality, they would become stateless persons. However, according to the 1961 Convention on the Reduction of Statelessness and the 1954 Convention relating to the Status of Stateless Persons, governments are required to offer stateless persons residing in their country opportunities for naturalization. It is to be hoped that it will not come to that.

It is conceivable that countries affected by territorial loss would continue to be recognized even if their territory no longer exists. De facto states that have come into existence by violating international law are not recognized by the international community. Independence efforts that challenge the territorial integrity of a state are highly controversial. This applies to secessionist movements such as in Catalonia or the recognition of Kosovo. Countries that are created by acts of war, for example, Northern Cyprus, are generally not recognized and are not represented in UN bodies [51]. Conversely, states that disappear because of rights violations could still be recognized [52]. Moreover, cases exist in which governments have been recognized internationally, even though they had lost power. For moral reasons, the recognition of states which lost territory would be required, especially by industrialized states that have contributed to the loss of statehood through their emissions. It is only questionable how long such a status could be maintained de facto.

Another theoretical possibility would be to transfer the original national territory to the territory of another state and relocate the entire population there. To do this, however, countries would have to agree to give up parts of their territory in favor of the dwindling states or to lease them on a longer-term basis. While the first option may be unlikely, the second in turn raises questions of legal jurisdiction between the lessee and the host state [53]. In addition, large-scale state-led relocations often cause substantial problems for the affected communities.

Even if the territory of a state can no longer be inhabited, the population has a right to citizenship. Article 15 of the Universal Declaration of Human Rights states:

> 1. Everyone has the right to a nationality. 2. No one shall be arbitrarily deprived of his nationality nor denied the right to change his nationality [54].

Whether states would accept people who are forced to leave their country due to severe climate impacts and possibly offer them citizenship is an open question. Some scientists argue that even without a territory, a nation and its citizenship can continue to exist, even if it is spread over several places or countries. University of Hawaii law professor Maxine Burkett, for example, advocates a "nation ex situ" model in which affected persons retain their own citizenship and also adopt a second citizenship of the country in which they live [55]. The benefits of such an arrangement, Burkett argues, would be that the group's cultural identity and sense of belonging would be solidified through shared citizenship. But whether such a "nation ex situ" would endure for generations is rather doubtful under the current system of international law. Nevertheless, far-reaching reforms of international law could be initiated in the coming decades if the consequences of climate change increasingly put existing law to the test.

In short, existing law is not made for a world with a rapidly changing climate. The upheavals that lie ahead in a rapidly changing Earth system will challenge our legal system. Meeting this challenge will require proactive action, a willingness to change, and progressive approaches to interpreting existing law. Only in recent decades has humanity begun to rethink and scientifically comprehend the impacts of climate change. Legal change, on the other hand, takes time. Legal scholars and lawyers around the world are confronting the complexities of the developments and, through strategic litigation, new definitions, and legal change, are also adopting unconventional approaches.

Thesis

The existing border regime will lead to an increase in human rights violations in the wake of dramatic climate impacts. Unevenly distributed environmental harm requires institutional and legal reforms that open safe, legal migration routes from degraded areas and limit future damage.

Chapter 3
Islands Without a Future? The Disappearing Paradise of the Small Island States

Desolate Homeland: Survival in Paradise ♦ A Deck Full of Blood ♦ Superyachts ♦ Atoll Countries in Between Global Powers ♦ Expulsion from Bikini and the US Nuclear Legacy ♦ One Point Five to Stay Alive ♦ Planned Relocations in Fiji ♦ In the Eye of Irma: Superstorms in the Caribbean ♦ Evacuation at Gunpoint ♦ Billionaires and Homelessness: Who Owns the Islands? ♦ Toxic Algae Belt ♦ Grab-and-Go Capitalism ♦ Parallel Crises: Welcome to the Anthropocene

That Was it The thought crosses my mind as a freak wave catches our sailing catamaran sideways. I lie below deck in one of the two hulls of the "Okeanos Marshall Islands" and am tossed to the side by the force of the impact. Drowsy, I search for a hatch, climb on deck, and clip myself in. Our life jackets have a solid safety mechanism—a rope with a large carabiner that you hook onto steel cables, to move around even in stormy conditions. It is the middle of the night. Pouring rain is hitting us in the face. The crew is on their feet, fighting with all they have against the storm whistling over us. The mood is tense, but the team—experienced sailors from Tahiti, Fiji, and the Marshall Islands—is well attuned and knows what to do. Our sailing trip takes us from Majuro, the capital of the Marshall Islands, to Ailuk, a remote atoll of the country. We are traveling by sailboat for about 2 days. The "Okeanos Marshall Islands" transports items of daily use, a mechanical sewing machine, suitcases, and food such as rice. I am riding along into the storm to conduct on-site interviews with people whose relatives have migrated.

From Majuro's blue lagoon the day before, it was hard to imagine the forces which the ocean unleashes. The deep dark night makes nature seem even more powerful. Water masses surround the boat from all sides. Contrary to my fear that the sea would swallow us any moment, the catamaran bravely dashes over the huge wave crests that roll toward the sailboat like a black wall. It manages to emerge every time. After what feels like an eternity, the wind finally dies down, and morning breaks. The Pacific, which is considered unpredictable among sailors, commands my respect after this night. At dawn, I feel a deep connection to the ocean:

The sunrise is just as unforgettable as the nighttime experience; the world seems to rip open and expose its soul with a red-pink glow. Someone starts playing the ukulele; there is singing. The wild night is washed away as if it had never existed. When we finally anchor off the coast of Ailuk, the mayor of the island approaches us together with a small group on a boat docking to the catamaran. We cross over and, relieved, feel land under our feet again.

The Marshall Islands are an independent country located in the middle of the Pacific Ocean (Fig. 3.1). The outer islands still embody the image of a South Seas paradise: palm trees and white sand beaches littered with coconuts. Diving into the turquoise sea, one can observe friendly reef sharks, fat groupers, and orange-white clownfish. But the appearance of a supposedly perfect island world quickly erodes when residents describe their everyday lives. Older people tell me about the changes that are taking place on the islands as a result of climate change. A 77-year-old man reports: "It is much hotter than when I was young. The droughts are more frequent and last longer. These islands resemble a wasteland. The soil is poor. Now we hardly have anything to eat because there is not enough breadfruit and fruit. There used to be enough on the table." There is also a shortage of labor due to the exodus of younger people, which makes the situation even worse. One woman agrees: "Life has become harder." When talking to young Marshallese, the focus is not on "the past"; they are looking to the future. Inhabitants are fighting an inner battle between the desire to stay and the realization that there may no longer be a future for them, here in their homeland. With resignation, they realize what has already been lost. Many see hardly any more livelihood opportunities for themselves and their families on the islands.

Desolate Homeland: Survival in Paradise

There is a difference of 2494 feet (736 m) between the United States and the Marshall Islands. It is 2494 potentially vital feet, because the United States is on average 2500 feet above sea level and the Marshall Islands about 6. In times of global warming, which is accompanied by powerful storm surges and floods, this measure could define the future of an entire nation. The Marshall Islands, home to fewer than 75,000 people, are among the group of shallow coral atolls whose very existence is threatened by rising sea levels. They also include, for example, Kiribati and Tokelau, also located in the Pacific, or the Maldives in the Indian Ocean. Most of the population of the Marshall Islands lives on the main island. The nation has its own language and culture. But more and more people are migrating from the outer islands, some because they hope to get better educational opportunities, while others because the land can no longer support their families. On Ailuk, I talk to families who are planning to emigrate.

A young couple with a toddler welcomes us into their home, a typical island shack with a corrugated roof. Sheets cover the cracked concrete walls. Kilok and his wife Tracy are in their late twenties. While Tracy is busy with their child, Kilok does

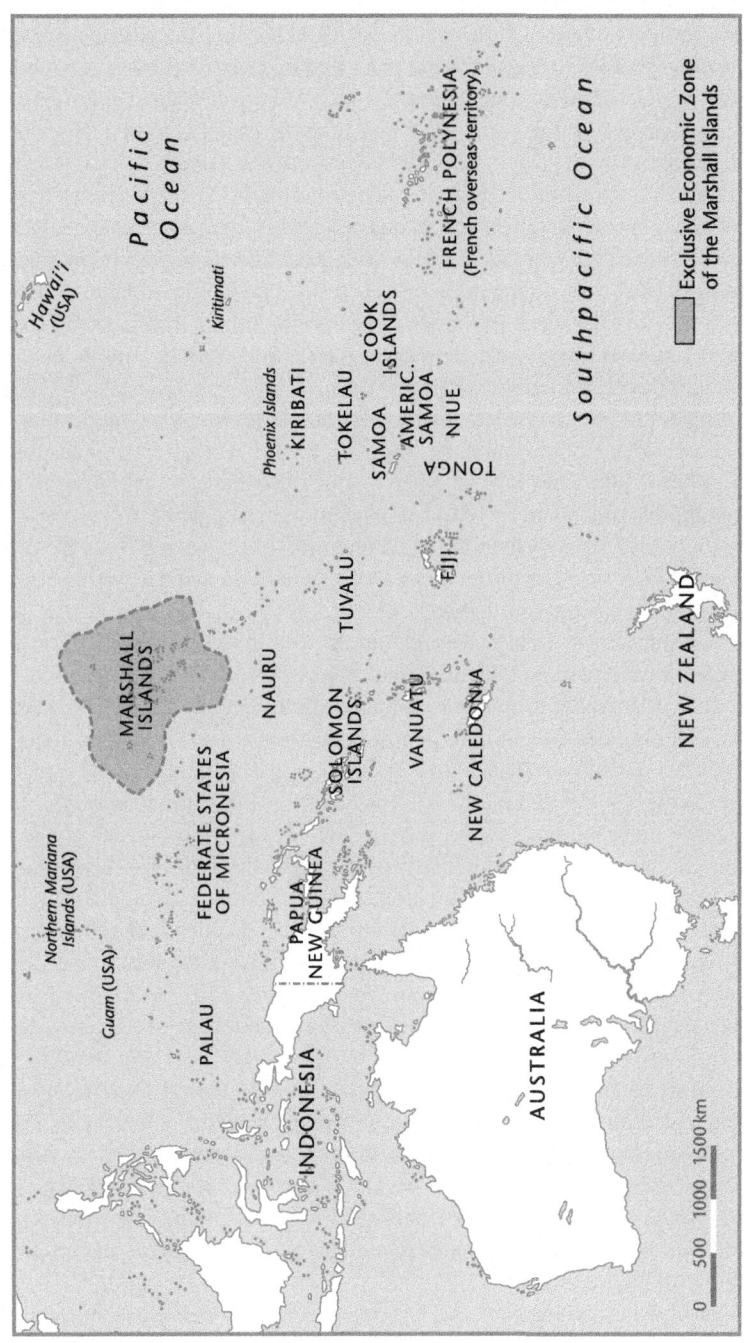

Fig. 3.1 Map of the South Pacific Region highlighting the approximate Exclusive Economic Zone of the Marshall Islands

most of the talking [56]. Both look exhausted. The air is stuffy, with the heat almost unbearable. The room stores the heat instead of dissipating it to the outside. I hardly see any traditional wooden buildings or palm roofs because the island's trees have become scarce. Therefore, cement is the main building material, although there is a lack of freshwater and also sand. The willy-nilly used salt water, and the tropical weather cause cracks in the walls. The buildings' lifespan is short; they quickly begin to crumble.

We sit on the floor while the couple talks about their life on the remote island. Neither of them speaks English; the manager of the "Okeanos Marshall Islands" translates the Marshallese for me: There is no Internet and no regular telephone connection; news is received via the radio or by satellite phone. The village community is tightly knit. The men work primarily in the production of copra, dried coconut meat, which is picked up by a ship at irregular intervals, processed into flour and oil in Majuro, and then exported worldwide.

Extracting copra from a coconut shell with a machete requires skilled hands. It is the main source of income for most families, but the price is so low at times that the government has to subsidize it. And when one of the freighters that picks up copra and delivers goods is delayed by weeks or even months, medicine and food become scarce on the island. The women make handicrafts, finely woven mats from palm leaves, braided necklaces, or turtles from shells, which occasionally still serve as a substitute currency when cash runs out.

Kilok and Tracy enjoy living on Ailuk and are proud of their culture. Their customs are closely interwoven with the land, the possibility to go fishing, to grow coconut palms and pandanus trees, or to obtain traditional medicine from endemic plants. As early as 1000 years ago, people settled here and developed their own forms of art and techniques for building sailing boats. All of this is now threatened by climate change. When talking to Kilok and Tracy, I sense how stressful it is to live with these constant worries. As soon as the conversation touches challenges, there is heavy silence. Finally, Kilok tells us about the difficulty for the young generation in finding fertile land for cultivation. Frequent floods salinate the soil, harming palm trees and other trees. "More than 50 of my trees (coconut palm trees) are damaged by the tides. They are my food crops. [...] We have less to eat now," a teacher on Ailuk already shared their concern with us.

On the islands, the impacts of climate change cannot be overlooked. Nature which has allowed people to thrive and survive on the islands for thousands of years is falling out of balance. Kilok looks anxiously into the future; he does not know why floods and droughts are becoming more frequent. Other villagers report that they are catching fewer and fewer fish in the shallow waters around the island—because of coral bleaching. Fishermen have to go further out to sea to make a good catch. This poses dangers, costs time and fuel, which has to be imported. Against the backdrop of the increasingly noticeable changes, Kilok sees no way to provide for his family in the long term.

Like many of his compatriots, he is now determined to leave Ailuk, the atoll where he has spent his entire life. His face reflects the responsibility weighing on him. The family plans to relocate to Oklahoma, United States, where they know someone who can supposedly find Kilok a job. Marshallese can immigrate to the United States without a visa because the country has a Compact of Free Association (COFA) agreement with the United States. The United States administered the Trust Territory of the Pacific Islands, a United Nations trust territory that was established over several Micronesian nations after the Pacific War, including the Marshall Islands. They tested atomic bombs and built a missile station on Kwajalein. The COFA agreement is a result of this history and was renewed in 2023 for another 20 years.

The exodus of the Marshallese today often leads first to the main island of Majuro, from there to the United States, and then mostly to Hawaii, California, Arkansas, Missouri, or Oklahoma. In these states, there is a Marshallese diaspora that often supports the arrival of their countrymen. Nevertheless, frequently, only low-skilled and badly paid jobs on the assembly line await them. In Arkansas, for example, many former islanders work in industrial chicken factories. The work in such a factory and the day-to-day routines stand in stark contrast to the life on a remote island and thus often comes as a great shock to the migrants. Moving to the urbanized capital Majuro or to the United States brings a radical change in their lives because their traditions, which are deeply interwoven with the land and the ocean, can hardly be adapted to the city environment.

But there is also another kind of migration from the Marshall Islands: Elites can afford to send their children to study abroad, for instance, in the United States where they often stay. Such migration often leads to a success story, a good education that protects them from a low-paying job on the assembly line. This opportunity is not within reach for Kilok and his family; their future is uncertain. In our conversation, I want to say to him and Tracey: Do not go to Oklahoma! But what would be the alternative? There are no easy solutions. I remain silent. We bid farewell.

Many migrants hope to return to "their island," even if they earn a bit more money on Majuro. Despite the difficult and increasingly risky situation, many want to stay on their ancestral islands as long as they can. A Marshallese politician reports what women have told her: "One of them said, 'I am not afraid of sinking in the sea. I'm afraid of moving away and being nobody,' and, 'If I migrate to the U.S. - who would I be?' They don't have coconut trees! They don't have pandanus trees! They don't have our fish! They have theirs, but it doesn't taste like ours. I would, ... I won't have an identity." The women are aware that migration ultimately leads to a loss of identity—a portent of the impending demise of an entire culture.

A Deck Full of Blood

The "Okeanos Marshall Islands" brings us back to Majuro. The sailing catamaran is all about sustainability. In addition to using the magnificent winds, there is an engine powered by coconut oil. The additional propulsion is used in adverse winds, or when entering a harbor or narrow bay. Solar panels generate electricity for the nautical equipment on board. The Okeanos features a mix of high tech and tradition, as the basic design of the New Zealand-built catamarans comes from centuries of Pacific boatbuilding [57].

During the trip, we are well supplied with rice, coconuts, and the occasional bowl of cornflakes. Crew members throw out bait, and then suddenly, one of them pulls a tuna on board. The silvery animal is huge, weighing more than me. It is killed and gutted with a knife. The crew's supply of fish is thus secured for the next few days, but the sight of the large animal carcass is intense. Blood splatters on the white deck and slowly flows into the sea. What does the death of an animal mean? Many people ask themselves this question when it comes to meat consumption in everyday life—but a fish rarely arouses pity.

Of course, it makes a difference whether people catch only as many fish as they need to survive, or whether huge fishing vessels equipped with trawls, radar, and acoustic sensors clear the seabed for industrial purposes [58]. In the face of this overwhelming force of equipment, there is no escape for many animals, not even for the tuna, which can dive to depths of up to 1000 m. Since the consumption of fish became globalized in the 1980s, stocks have declined rapidly, although there are signs of recovery in some areas. Protection zones are, however, often not sufficient for species that move over long distances and thus also reach less regulated territories. Catch quotas are set too high *and* not adhered to. Often, the fines for violations are simply part of the companies' calculation, as tuna in particular fetches very high prices. Many of the animals that are auctioned today at large fish markets, such as Toyosu in Tokyo, do not even reach reproductive maturity. This means they cannot reproduce before they are killed, leading to further depletion of stocks. Until the last tuna ends up on someone's plate. But it is not just about the tuna. All kinds of species perish in the trawl nets as bycatch. The nets are also causing massive damage to the seabed, including by releasing CO_2 that is trapped there.

Smaller fishing boats anchor in the harbor of Majuro. Not far from the quay walls is a tuna factory, Pan Pacific Foods Inc., which processes the fish for export. Talking to some of the employees, it quickly becomes clear that they are happy to have found a job at all. In addition to a few retail stores, a handful of hotels, and small businesses, one of the larger private companies in Majuro is "Tobolar," which processes dried copra into flour and oil. Copra flour is supplied as animal feed mainly to Australia. The oil is used as a raw material for a whole range of products, from cosmetics to medicines or even fuel. In addition to the factory, employment opportunities are provided mainly by state and local authorities.

Nutrition is a worry for many. In urban centers like Majuro and Ebeye, traditional self-sufficiency through small-scale agriculture and fishery no longer works. The population is dependent on food imports, and these need to be paid for. Only few people can regularly afford fresh fruit and vegetables; most subsist on highly processed foods or cheap rice with chicken. But malnutrition has consequences: More than half of the adults suffer from obesity, and the population has one of the highest rates of type 2 diabetes in the world.

Superyachts

Shortly before Christmas, we want to voyage with the "Okeanos Marshall Islands" to Maloelap, an atoll west of Majuro in the Central Pacific. People on the atoll have been waiting for food aid for some time, which had arrived on the main island during a severe drought driven by El Niño. Meanwhile, my father Hermann Vinke has landed in the Marshall Islands. As a foreign correspondent who already traveled to Micronesia multiple times in the 1980s, he decided to accompany me for a while on this special expedition. During his work in East Asia and the Pacific, the focus was on the adverse consequences of the testing of nuclear bombs. Now, climate change poses the greatest risks to the island chains.

Before the crossing to Maloelap, we want to buy some extra provision for the journey. When we arrive at the supermarket, the shelves are almost empty. No fresh vegetables or fruit, no packaged food. With some mushy toast, we leave again. At the jetty, we meet a local who tells us that he visited all the supermarkets, and everywhere the same picture: empty shelves. He also seems to know the reason. The crew of some of the yachts anchored in the harbor of Majuro had supposedly stocked up with the food supply of the main island, in order to be able to celebrate the Christmas days near Ailinglaplap, a remote island with a luxury resort. The crew was perhaps not interested in the fact that now, shortly before Christmas, hardly any fresh food is available on the islands. One of the vessels anchored in the harbour is Google founder Larry Page's yacht "Senses". It is equipped with everything else a multibillionaire could desire: surfboards, jet skis, and a helicopter.

I experience one of the most surreal scenes on my trip when loud electronic music suddenly blares from one of the yachts over the harbor area, while the crew dances on deck in skimpy clothing like in a television commercial. Next to the "Okeanos" which we are stocking with the bags of food aid, a run-down fishing boat is anchored with Burmese sailors who, to all appearances, have not been ashore for some time. In their neighborhood, ships anchor that resemble rusting wrecks. All this does not dampen the mood on board.

Atolls in Between World Powers

The journey to Maloelap is again accompanied by stormy weather. But the sight of the lagoon washes away any signs of strain. A few crew members and I jump into the sea with mask and snorkel. Colorful fish circle in the turquoise waters; huge corals still alive wobble in between, and then suddenly, the contours of a shipwreck become visible. It bears witness to the fierce battles between American and Japanese forces during World War II. Our Captain Alex sights a large shark, which we now want to see as well. My father worries that we will become surrogate food, but sharks rarely attack swimmers; rather, these wondrous animals die far too often at the hands of humans. But the shark moves faster than we do and swims away into the depths of the ocean.

We, on the other hand, still have some work ahead of us, because the cargo of the "Okeanos" needs to be unloaded. After we have brought the rice sacks ashore with our combined efforts, we get to talk to the residents of Maloelap. A local politician explains: "The people here are trying their best, but it is hard for them. The island's resources are exhausted, incomes are falling. Coconut harvests are declining; so are fish stocks. (...) Many people have left the island because of this. Even some of my own family migrated." Another agrees, "People can't support themselves here anymore. When food becomes scarce, they look for other options out there."

On land, there are also still remnants of the Pacific War. During World War II, Japan maintained a naval base with runways for planes that brought supplies. Tropical plants climb up the ruins of a multistory building that was once the headquarters. Large holes, eaten by rust, gape in the huge tanks that supplied fuel to the warships. Elsewhere, nature lays its lush greenery over a large bunker. A single antiaircraft gun rises from the shallow lagoon. Maloelap now calmly bears the traces of history. The islands and atolls in the Pacific have repeatedly been instrumentalized by great powers.

Outside influences go back a long time. In the second half of the nineteenth century, Christian missionaries came to the islands to convert inhabitants. They gradually displaced the local religion and culture. The missionaries had followed the merchants who had already begun to invest in coconut plantations and to boost the trade in copra. Leading German trading houses such as J. C. Godeffroy & Sohn in Hamburg profited from a buoyant trade in copra, mother-of-pearl, and tortoiseshell.

Before Japan occupied the Marshall Islands at the beginning of World War I, they were under German colonial rule from 1885 to 1914 and functioned as a provisioning and coal station for the Imperial Navy. Here, warships were reloaded with fuel and food. Even though formal treaties were signed with tribal leaders authorizing German activities on the islands, the people of Micronesia could hardly have rebelled against a military power like the German Empire. Because of this ramified history, many Marshallese also have German names, such as the president of the Marshall Islands, Hilda Heine.

When we return to Majuro, the superyacht "Senses" appears somehow changed, and I ask our captain if the luxury ship has turned around. "No," he replies. "That's the other Google yacht anchored here." The "Dragonfly," at over 70 meters long, surpasses the "Senses" and belongs to Google's second founder, Sergey Brin.

On average, the richest 1% of humanity produces more than 30 times as many greenhouse gases per capita with their private jets, yachts, and other consumer goods as is compatible with the 1.5 °C limit [59]. Yet, the share of global emissions attributable to the lifestyles of the superrich continues to grow and could reach 16% by 2030. Inequality, climate change, and malnutrition—the world's problems—are emerging from the Marshall Islands as if under a burning glass.

Problems are also piling up in the capital Majuro. The densely populated atoll, which rises like a snake from the Pacific, has few beaches left. Sand is dredged away to pour concrete, and sea walls, which are supposed to protect infrastructure, ultimately contribute to the slow erosion of the fine sand off the coast. There is still a small stretch of beach in front of the campus of the University of the South Pacific. A refrigerator is rusting away in the sand—obviously washed up. The waves crash against it. Everywhere on the island, garbage is a problem. There is no waste treatment plant, and more and more plastic is arriving on the islands together with imported goods from the People's Republic of China. The highest point of the island is now a landfill on Majuro Atoll, which is constantly growing and jokingly called "Trash Mountain".

An employee of the city administration reports a plan to blow up a reef in order to sink a part of the garbage there and then sealing it with a concrete shell. This structure would then double as a breakwater. "It's a dead reef," the man says, "we drill it, then we put dynamite in it and blow it up." Incredulous, I look at him, hoping this could be a joke. But I quickly realize that in a place where the world's problems seem to cluster, solutions are desperately sought. "We don't have the technical skills and knowledge to deal with the consequences of climate change and various environmental stresses," he emphasizes.

In many places, discarded cars blemish the environment. Cars in general are another problem on Majuro. There are too many of them; even traffic jams form on the one main road. The disposal of the discarded vehicles is not regulated. "We have copied the American lifestyle here," says my interlocutor, shrugging his shoulders.

The Expulsion from Bikini and the Nuclear Legacy of the US

On a Sunday in 1946, the American military governor Ben Wyatt gathered the inhabitants of Bikini Atoll in the Marshall Islands after church service and reminded them of "the Children of Israel whom the Lord had saved from their enemy and led into the promised land" [60]. Like this biblical exodus, the inhabitants were to leave Bikini: "for the good of mankind and to end all world wars [61]" was the flimsy justification. After the end of the World War II, the United States wanted to test atomic bombs in a remote area.

The missionized Marshallese on Bikini reluctantly agreed to a temporary (!) resettlement—what else could they do faced with American military might—and were brought to the atoll Rongerik. Most of them hoped to return soon. However, Rongerik was much smaller than Bikini and did not have enough fishing grounds to feed the resettled people. Months of starvation followed. Because the American occupiers did not provide enough food for the population, the situation grew worse and worse. Now the men, women, and children were shipped to Kwajalein and later again to Kili, where some of them still live today—dependent on food imports because Kili has no lagoon. Thus, there are little possibilities for fishing.

In the years from 1946 to 1958, the United States undertook a total of 67 nuclear bomb tests on Bikini and Eniwetok. The test series culminated in "Operation Castle," when thermonuclear bombs of enormous explosive power were detonated. Disaster was inevitable. In 1954, the hydrogen bomb "Bravo" exploded, which was 1000 times more powerful than the atomic bomb that devastated Hiroshima in World War II. The radius of the radioactive ash-like fallout was much larger than originally thought. Residents on nearby Rongelap Atoll and fishermen out at sea were exposed to the nuclear ash fallout and soon became afflicted with radiation sickness, some of them severely. It has been stipulated that their exposure to radiation was done deliberately in order to later examine its effects on humans.

On Ailuk, I spoke with a surviving fisherman who saw the fireball, the glaring light of the explosion with his own eyes. He did not know what it was. Many of his friends later died of the consequences, mostly of cancer. He himself was lucky and survived, continued to work as a fisherman, and spent his life on Ailuk. To this day, the nuclear tests of the 1940s and 1950s mean a heavy burden for the Marshall Islands. Entire islands evaporated under the atomic blasts. Many residents still suffer from the nuclear contamination of their environment. The number of cancer patients is high, and deformities in children due to late effects of radiation damage are not uncommon, while health care is inadequate.

Before the people of Bikini Atoll became nomads of the nuclear age, they had lived for thousands of years according to the unwritten rules and laws of nature. The environment provided them with enough food, an abundance of fruits and fish. With the United States' trusteeship over Micronesia, many things changed. A Coca-Cola culture took hold, undermined the traditional life on the islands, and lured especially young people with false promises, which eventually led to a complete dependence on the United States.

Decades passed until in the late 1970s, the US government finally started collecting the nuclear waste resulting from the test series, sinking it into a crater created by one of the explosions, and then sealing it with a concrete blanket. The cleanup operation was superficial. In particular, it involved storing material contaminated with plutonium. Plutonium has a half-life of 24,000 years.

A large part of the residues also ended up in the Eniwetok lagoon. The population and environmental experts protested strongly against this, but in vain. The conditions for the construction of the nuclear silo were lax and stood in stark contrast to the dangers posed by the "Runit Dome": "The dome does not even meet the standards for landfills for American household waste," says Prof. Michael Gerrard, who founded the Sabin Center for Climate Change and Law at Columbia University [62]. He said the radioactivity of the landfills was much higher than approved limits because of inadequate safeguards.

For several years now, the "Runit Dome" has been showing cracks, which means that the concrete shell is becoming porous. Weathering and lack of maintenance are taking their toll on the material. As sea levels rise, the nuclear storage facility is likely to be flooded one day. A tropical cyclone could also rupture the concrete cover, releasing highly radioactive materials into the sea. The US authorities, however, see no reason for concern. They say that the lagoon is already so heavily contaminated with nuclear material that no further potential pollution is considered significant.

Today, the history of displacement is repeating itself, also caused by environmental pollution and destruction by powerful industrialized nations like the United States. Emissions that other countries release into the atmosphere hit low-lying island nations in the Pacific, such as the Marshall Islands, particularly hard. Only this time, the displaced cannot find protection within their own country. The entire territory of the state is affected.

One Point Five to Stay Alive

The outlook for the small island states is grim. If the Earth warms by more than 1.5 °C (today, we are already at 1.2 °C above pre-industrial temperature levels and have already had a 12-month period between 2023 and 2024 that was warmer than 1.5 °C above pre-industrial temperature limit), flat-lying atolls could become practically uninhabitable in the long term, due to sea level rise and the combined effects of storm surges and freshwater scarcity. When I am asked what makes the difference between 1.5° and 2 °C warming, I often refer to the existence of these small island states. Sometimes, the response I get is another question with a certain undertone like: How many people even live there? As if some states and cultures were dispensable, as if they were the price to be paid for the fossil growth of the global middle and upper classes. Some people would rather close their eyes to the inhumane consequences behind this logic. But already today, the damage to people and nature caused by climate change is undeniable.

Scientists predict that people will be forced to migrate from shallow atolls first due to salinization of freshwater and later also due to land loss. Atolls do not have deep groundwater deposits, but small so-called freshwater lenses. When a storm rages, the rise in sea level means that larger and larger areas are flooded—with salt water. If this causes the small freshwater resources to become saline, people become dependent on rain. If, however, the rain fails to fall and the number and intensity of droughts increase at the same time, the problems can hardly be overcome. For this reason, some scientists assume that many atolls could become uninhabitable by the middle of the twenty-first century, even if sufficient land mass would still exists [63].

Politicians in the small island states are faced with difficult decisions: How to fight to preserve their homeland by exerting pressure at the international level to reduce emissions and whether to explore how parts of the population can migrate early and under tolerable conditions. For example, the former president of Kiribati, Anote Tong, tried to establish the concept of "migration with dignity," and in this context, his government bought land on Vanua Levu, the second largest island in Fiji. People from Kiribati should be able to settle there, because the situation on the coral atoll is even more tense than in the Marshall Islands: A lower per capita income of the inhabitants limits their possibilities for adaptation, and the population is growing. However, while his passionate call for more ambitious climate policies has been widely recognized in international forums, Tong's forward-looking policies have been met with outrage by some of his countrymen. He was accused of being ready to abandon his country.

A similar pushback occurred after the government of Tuvalu signed a treaty with Australia, the "Falepili Union," which committed the two countries to jointly build climate resilience and would enable Tuvaluans to migrate in dignity to Australia as part of a labor migration scheme. Moreover, Australia and Tuvalu would deepen their security collaboration, which serves Australia's interest in light of Chinese expansionism. Tuvalu's then prime minister Kausea Natano, who sealed the deal, did not hold consultations with civil society organizations and lost the election a few months later. The treaty was a major issue in the election, as opponents suggested it would curtail Tuvalu's sovereignty. In exchange for the migration scheme, Tuvalu accepted to only enter into security arrangements with other countries if Australia agrees.

At a side event on the sidelines of negotiations at the 2021 global climate conference (COP26) in Glasgow, Scotland, I was part of a panel on "Climate Change and Displacement," as was a delegate from Kiribati, Teea Tia. She addressed the audience with the question, "Why should our countrymen migrate, leave the islands and their way of life behind because of climate change?" The question "Why us?" struck at the heart of climate injustice, as it underscored what the global community is willing to impose on countries like Kiribati. The Pacific Climate Warriors, an alliance of mostly young climate activists from Pacific Island nations, unite behind the cry "We are not drowning, we are fighting!" They, too, are determined to fight for the protection of their homeland. The right to stay and the possibility to migrate are, however, not mutually exclusive. Both options must be equally available to those affected in light of the looming dangers.

Several governments have repeatedly put the climate issue on the Security Council's agenda. Although no resolution was passed—the veto powers of the United States under Donald Trump, Russia, and China would have blocked this—the connection between climate impacts and security is now being taken up in more and more international forums. The German government, i.e., is funding the first climate security adviser to a UN peacekeeping mission, Briton Christophe Hodder. In Somalia, he shares his expertise on environmental issues that also affect the security situation in the country.

In addition, the Federal Republic of Germany founded the Group of Friends on Climate and Security together with the Pacific small island state of Nauru. The group aims to anchor the issue in various UN bodies, such as the Security Council, and to engage in agenda setting. Though the Biden administration joined the Group of Friends, another attempt to pass a resolution in the UN Security Council on climate change failed in 2021 due to Russia's veto. India, which has a nonpermanent seat on the Security Council, also voted against the resolution, which was introduced by Ireland and Niger. China, on the other hand, abstained. Of the 15 members of the Security Council, 12 countries voted in favor.

There may be many reasons why a country like India, which is extremely affected by climate impacts, voted against the resolution. For one thing, part of the political elite is closely tied to the coal industry and wants to continue using fossil industries until the middle of the second half of the century [64]. For another, states such as Russia want to maintain a traditional and narrowly interpreted notion of security for the Council so that national sovereignty is not eroded by referring domestic nontraditional security risks to the Security Council. This position may have been echoed by India. The justification given was that the resolution would undermine the process of international climate change negotiations taking place under the United Nations Framework Convention on Climate Change (UNFCCC). However, this hardly seems plausible. The resolution simply aimed to include the role of climate impacts in addressing conflict and counterterrorism. Its failure is another missed opportunity to battle climate risks through multilateral action at the highest political level.

In Service of Humanity: Tony de Brum's Legacy
Former Marshall Islands Foreign Minister Tony de Brum (1945–2017) and the High Ambition Coalition he founded were instrumental in anchoring the 1.5 °C limit in the Paris Climate Agreement. The charismatic politician campaigned for the higher level of protection with the slogan "1.5 to stay alive." Early in his political career, he was concerned with man-made environmental problems. As a young boy, he was an eyewitness to the 1954 "Bravo" atomic bomb test on Bikini Atoll; later, he called for a ban on nuclear weapons during his time in office, as well as adequate compensation for compatriots contaminated by the tests. Finally, he turned his attention to addressing the climate crisis.

With diplomatic skill, he won over governments of partner countries for his interests in the negotiations on the Paris Agreement, such as the German government, which supported the concerns of the small island states. The Climate Vulnerable Forum [65] founded by the Maldives and the Alliance of Small Island States (AOSIS) [66] also joined the initiative. And they were successful. Article 2 of the agreement sets a target that "the increase in the average temperature of the Earth shall be kept well below 2 °C above pre-industrial levels and efforts shall be made to limit the temperature increase to 1.5 °C above pre-industrial levels, recognizing that this would significantly reduce the risks and impacts of climate change." [67] Thus, the 1.5 °C limit became part of the international agreement.

However, this agreement is not only of great importance for the small island states. Fighting for the 1.5 °C limit ultimately means a service to humanity. After the Paris Climate Agreement, the Intergovernmental Panel on Climate Change presented the 1.5 °C report and showed how disproportionately higher the damage to the environment and natural livelihoods would be under a 2 °C scenario [68]. For all countries, the risks would again be significantly exacerbated, for example, by the higher probability of severe heat waves occurring.

In 2015, Tony de Brum received the Right Livelihood Award, also known as the Alternative Nobel Prize, for his commitment to the fight against climate change and for nuclear disarmament. At the time of the negotiations in Paris and the award ceremony, Tony de Brum was already seriously ill. He died in 2017, and his voice is since sorely absent from international climate change negotiations, but his political legacy endures. Vulnerable nations are increasingly vehement in demanding emissions reductions. The slogan "1.5 to stay alive" was flipped in the run-up to COP26 in Glasgow 2021 to read "keeping 1.5 alive." The ambition for the summit was to keep the 1.5 °C limit within reach.

In the face of rapidly rising emissions, this is no easy task. Nevertheless, nations were at least able to agree on a clear affirmation of the temperature limit in the final document of the Glasgow conference, backed up with timely emissions reduction targets. In the Framework Convention on Climate Change, the international community recognizes "that limiting global warming to 1.5 °C requires rapid, deep and sustained reductions in global greenhouse gas emissions, including a 45% reduction in global carbon dioxide emissions by 2030 compared to 2010 levels and a net zero reduction by mid-century, as well as deep reductions in other greenhouse gases." [69] The global stocktake of 2023 also reflected that countries were not on track to achieve the necessary emissions reductions. The cover decision of the COP28 in Dubai was the first to include explicitly language on "transitioning away from fossil fuels." It is a seemingly weak achievement, since one would assume that all prior 27 negotiating rounds must have had considered the reduction of fossil fuel use. But at the climate summits, political objectives are negotiated and formulated that are based on compromises between very different states. The small island states, whose existence depends on far-reaching climate protection, must ultimately come to an agreement with the oil states, whose economic model is based on the fossil fuel system. The fact that this multilateral balancing act is producing an increasing level of

> ambition at all—albeit very slowly—is an achievement in itself. Nevertheless, this process, which began with COP1 in Berlin under the auspices of then German Environment Minister Angela Merkel in 1995, is not enough to turn the tide toward sustainability. Rather, a multi-level approach is needed in which civil society groups, cities, municipalities, and regions provide new impetus for nation-state action so that compromise at the intergovernmental level is not the only guiding principle for action.
>
> In addition to targets and political promises, what is needed above all is more concrete implementation—also to strengthen people's trust in politics by holding governments accountable for the achievement of targets. This means that the second half of the 2020s must focus on implementing existing goals. This will require major transformations in all sectors of the economy, in public and private life. Tony de Brum has shown how much influence a single person and one of the smallest states in the world can exert.

Planned Relocations in Fiji

Not in all the island states is the situation similar to that of the flat-lying coral atolls, such as the Marshall Islands or Kiribati. Even within the Pacific region, geography varies strongly. Fiji, for example, has a territory which is about 100 times larger than that of the Marshall Islands. It has mountains and lush rainforests. Nevertheless, parts of Fiji's islands are also threatened; entire coastal villages have already been razed to the ground during tropical cyclones, and others are in danger due to rising sea levels and the accompanying coastal erosion.

In 2019, I had the opportunity to participate in a Talanoa, a village community dialogue forum, on Fiji's main island of Viti Levu in the village Qelekuro. The small village had been severely affected by rising sea levels and severe storms. The focus was on whether the local community should develop a plan for relocation to a safer place inland and, if so, where new housing could be built in the future. The Fiji government, with the support of the German Development Cooperation (GIZ), already published Planned Relocation Guidelines in 2018 and Standard Operating Procedures for Planned Relocation in 2023. Adaptation alone will probably not allow all previously settled areas to continue to be inhabited.

Just how difficult and emotional such discussions can be is evident in many areas affected by disasters. From the perspective of those affected, it is understandable that people want to rebuild where they lived before disaster struck. Only few want to resettle to a different area. But whether these spaces can be effectively protected in the future is uncertain; technological and financial solutions must be found. Examples of misconduct after disasters are plentiful, for instance, in Maui, Hawaii, wildfires destroyed the city of Lahaina, and land speculation ensued.

In view of the already critical situation in parts of Fiji, the country's Planned Relocation Guidelines reflect "that planned relocation within Fiji does represent an option of last resort, however it is expected to become a more common response to climate related events in the future." [70] Difficult times lie ahead for the island paradise, but some communities are already taking action and exploring their options—for example, in a Talanoa.

A traditional Talanoa like the one in Qelekuro begins with several rituals designed to demonstrate mutual respect among the various panelists and to create a friendly atmosphere. We hand over some crops to the village representatives as a sign of our gratitude for their hospitality. The discussion takes place in a relaxed setting and thus offers space to work out solutions together, even in the case of opposing positions. We meet a group of women, the mayor, and young villagers in a community house. My colleague Teddy Fong from the University of the South Pacific in Suva, Fiji, and I sit barefoot cross-legged on handwoven mats on the floor and drink kava, a beverage made from the roots of the kava plant, which is said to have a calming effect. It is traditionally served at ceremonies in the Pacific. In a modified form, it is even sold as a soft drink in some countries today. Even though I know the gray-brown beverage from my student days in Hawaii, it still has a peculiar taste for me. I sip thoughtfully from the bowl in which it is served and skip most refills, worried that I might miss an important detail of the conversation in the kava fog.

Different types of environmental damage are addressed in the discussion. Both local problems, such as overfishing, and climate change with its multilayered consequences are explained in hushed voices by the participants. The tropical cyclone "Winston," for example, damaged and destroyed a large part of the village's modest houses in 2016, and one woman lost her life [71]. Drinking water supplies were contaminated by flooding, resulting in the spread of typhoid fever. Some damage to infrastructure is still visible today: battered wooden walls and shakily rebuilt small protective walls of stones that could hardly withstand a major flood. Another storm of the strength of "Winston" would perhaps destroy the entire village. For this reason, the residents are faced with the question of whether to continue to invest in protective measures on site or whether the families will have to decide to relocate.

During the conversation, a generational divide emerges. While the younger generation is more willing to move further inland, the older generation wants to relocate only within the existing village boundaries. They fear a split in the village community. The pros and cons are exchanged. Because the fishing grounds are shrinking and the land around the village is becoming scarce, younger men are already working inland on fields that partly belong to the village. To do so, they walk long distances every day. The older inhabitants express understanding, but they worry that their traditions will be lost if the village has no future in its existing form. The decision is still pending, especially since there is also a lack of money to financially support the resettlement. A commission is to bring about a coordinated decision on the future of the village in the interest of the residents.

After the end of the Talanoa, the village headman leads us through the village and shows us the place where the water overflowed its banks and unleashed its destructive power. In view of the small buildings, it becomes apparent what living in such an exposed area may mean under future climate change scenarios. I also feel the inner conflict of the village leader, who wants to protect his community, but, at the same time, points out to us the cultural sites of the village: old graves and areas that are of religious significance no one wants to leave behind.

The Talanoa opened my eyes to the diverse social contexts in which climate change unfolds and how many villages, communities, and cities face new challenges for which solutions need to be found. But it also showed me that very different points of view can be exchanged in a calm and relaxed atmosphere. When considering the continuing loss of our discussion culture, hardened opinions, and the outrage overload in social media, I find the Talanoa an inspiration. If there were no such forum for negotiating opposing positions in the village, disagreements could quickly turn into tangible conflicts.

The format of the Talanoa was also established at COP23, which was hosted in Bonn in 2017 under the presidency of Fiji, as a new element of cooperation and negotiation at eye level. While Germany did not cover itself with glory as host and as logistical organizer—I still remember the numerous desperate delegates of different origins who searched in vain for information at Bonn's main train station—the Pacific spirit, however, revitalized the negotiations in the long run.

In the Eye of Irma: Superstorms in the Caribbean

In another corner of the world, on the Caribbean islands, superstorms threaten people's lives. Hurricane Irma left a trail of devastation in 2017 when it passed over various Caribbean states with maximum wind speeds of 290 kilometers per hour (180 mph). Barbuda in the eastern Caribbean was hit particularly hard. Irma's eye raged for hours on the small island, destroying the entire infrastructure. Miraculously, only one person died—tragically, a young child. Just 2 weeks later, another storm hit the region, Hurricane Jose. All residents had been forcibly evacuated before the second storm was supposed to arrive, but it largely spared the islands. The probability of storms hitting the islands in the Caribbean with the most violent force is increased by climate change (Fig. 3.2) [72].

Two years after the disaster, I visit Barbuda. I want to find out how reconstruction is going, whether people have been able to return and what political measures are needed to better manage evacuations and return processes in the future [73].

The journey to Barbuda turns out to be complicated. I first go to St. Lucia, a mountainous island that was the filming location for the blockbuster "Pirates of the Caribbean." Shortly thereafter, a tiny propeller plane takes me to Antigua. The noise is deafening, and the plane bounces up and down until we land safely on Antigua a short time later. The next day, we are to sail from here to the neighboring island of Barbuda.

Although Antigua and Barbuda form a common state, the differences between the two islands are enormous. Ninety-seven percent of the total of almost 100,000 inhabitants live on Antigua. Because of this great imbalance in population

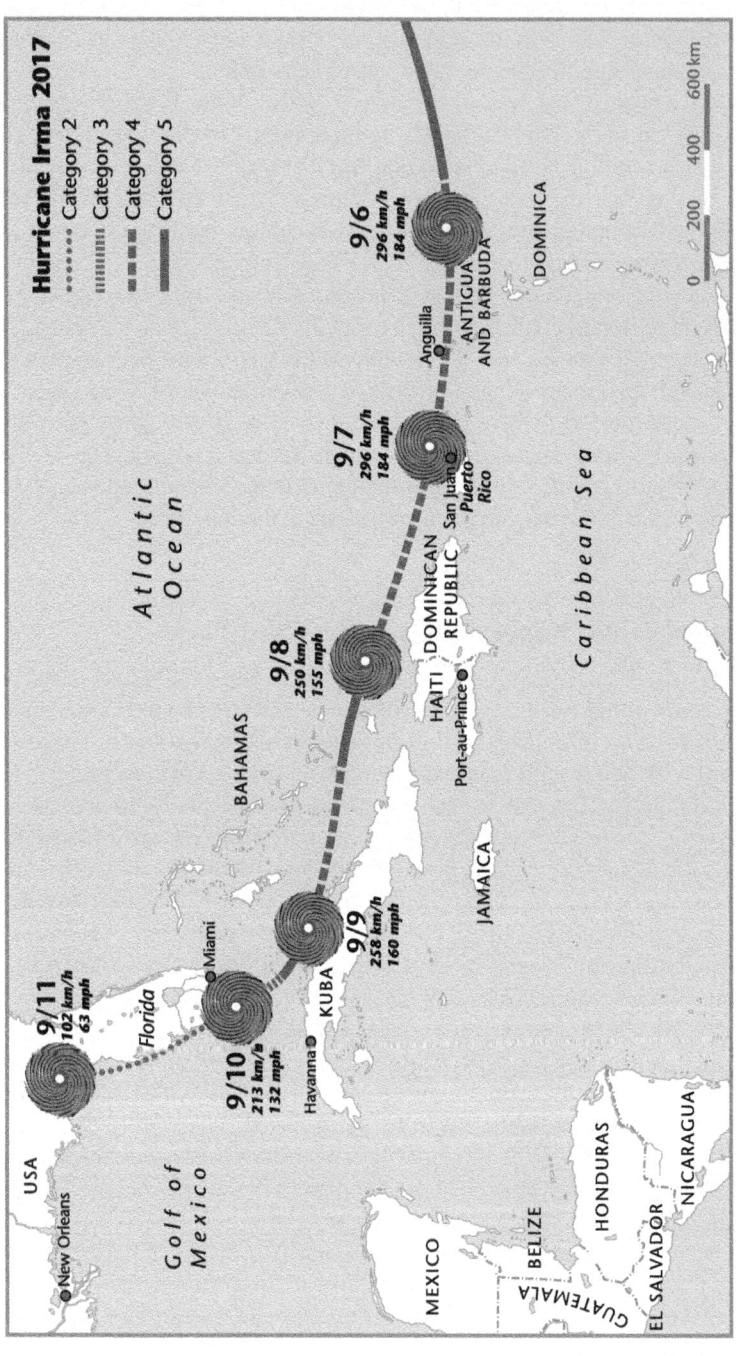

Fig. 3.2 In September 2017, Hurricane Irma passed over the Caribbean and parts of Florida in the United States. It left a trail of destruction. The map shows the storm tracks schematically

distribution, Barbuda has virtually no influence over the national government. Barbuda has only one member of parliament. The islands also differ greatly economically. Antigua has become an anchor point for Caribbean cruises, with masses of US and British passengers pouring into the port daily to visit the numerous souvenir stores and restaurants. Cruise tourism on Barbuda, on the other hand, plays a marginal role.

The history of the twin state of Antigua and Barbuda is marked by colonization and the slave trade. In the seventeenth century, enslaved people from the African continent were brought there to grow sugar for export to Europe under British colonial rule. The food cultivated on Barbuda was mainly used to feed workers on Antigua's sugarcane plantations.

As I walk through the colorful market in St. Johns, the capital of Antigua and Barbuda, I have an oppressive feeling in this strange environment. Today, shot glasses with a Caribbean motif and cheap shirts from China are for sale here. The houses are brightly painted. Tourists shop carefree and enjoy the sun and the Caribbean feeling. But the postcard scene is deceptive. It is not only the colonial heritage that weighs heavily on the island state, but also the consequences of climate change, which have widened the social divide between Barbuda and Antigua. I quickly walk on to the harbor and inquire when the ferry to Barbuda leaves. There is no daily connection at this time, and so I sleep one night in an empty hotel outside town.

When I arrive on Barbuda the next morning, even from the small ferry, the destruction caused by the hurricane is visible. Where there used to be palm trees, only brown stumps stick out of the ground. The wind whistles around my ears as I go ashore. My host Claire, who runs a small café on the island and rents rooms to travelers, is already waiting at the harbor. She helps me make contacts for interviews on Barbuda. As we cross the residential areas, it seems to me as if the storm had raged across the island only yesterday. Destroyed houses, rubble, and debris are everywhere. Water and electricity are sporadic. And that is 2 years after the disaster!

I meet a group of young men who lost everything in the hurricane. They describe their fear during the storm, the crashing of the shattering houses, and the roaring wind. In the course of the forced evacuation, some stayed with relatives in Antigua, while others had to hold out for months in emergency housing, like a sports stadium. They speak of traumatic events that are difficult to overcome. Their words are tinged with anger and disappointment. The young people have the feeling that they are being left alone to deal with the crisis. One of them wants to show me where he used to live. We cycle through the devastated streets to a property where only the bare concrete foundation can be seen. "This is where my house used to be," he says.

Some residents have received help from nongovernmental organizations and have been able to rebuild their homes. Others still live in tents. There is a lack of money and building materials to build new shelters. Antigua and Barbuda are heavily in debt, with national debt amounting to about 87% of the gross domestic product. Many small island states face similar problems. Due to ever more damage caused by climate change, it is becoming increasingly difficult for them to repay loans and raise new money.

Evacuation by Force of Arms

An elderly man working on his destroyed house not far from my accommodation tells me that he did not want to leave the island after the storm. But government troops forced him to leave at gunpoint, he says. The forced evacuation was another traumatic experience for many people on Barbuda. Although the second storm spared the island, the population was not allowed to return for a prolonged period of time, not even to make repairs to their homes. It was a fatal decision, because the weeks of absence increased the damage and rain and wind destroyed the last remaining personal belongings and made the houses completely uninhabitable. Quite a few residents had suspicions that the evacuation was not a purely precautionary measure of the government, which is based in Antigua.

Barbuda has long been under enormous pressure to follow the same development path as Antigua and open up more to tourism. But the vast majority of the island's inhabitants reject this. Barbuda has one distinctive feature: The people who live there share the island's land. This practice has a long history. After the withdrawal of British colonial rule, the previously enslaved people were left to fend for themselves on the resource-poor territory, but were able to establish a functional subsistence economy through communal farming. To this day, there is no individual land ownership. Any new construction projects or outside investments must be approved by Barbuda's elected Council. Thus, large parts of Barbuda have retained their natural beauty, while on most Caribbean coasts, oversized cruise ships obscure the view and pollute the environment.

Even after the storm disaster, communal use meant that people were able to provide themselves with the most basic necessities. But this practice is a thorn in the side of the national government, which seeks to privatize the land to attract foreign investors and boost the economy. The population opposes this, because potential investors can offer much higher prices for land than the traditional inhabitants and may then force them off their land in the long term.

This concern is not unfounded. When the inhabitants of Barbuda finally returned to their island after the evacuation, they found that their absence had been used to push ahead with another controversial project: the construction of an international airport. This airport is intended to provide access to the island for luxury tourists in particular and is therefore virtually the prerequisite for new resorts. According to the government, it is being financed by private investors. While the population was kept away from their ancestral land, old forests could be cut down in peace and quiet and archaeological sites seized in order to build a landing strip—all on behalf of the government and without the approval of the Barbuda Council. While lawsuits filed by local activists have halted construction for some time, the airport opened in October 2024.

Billionaires and Homelessness: Who Owns the Islands?

Not only in the Caribbean are there numerous examples of the disastrous influence of foreign investors on the housing market. In the Pacific, too, such as in Hawaii indigenous groups and long-established residents often can barely afford housing. The number of homeless people is rising. With thousands of homeless people, the 50th US state has one of the highest number per capita of homeless people in the United States. A disproportionate share of this group is from Micronesia [74]. Some have left their homes because of economic constraints, others because they desperately needed access to medical care, for example, due to diabetes or cancer. The increased incidence of these diseases is not purely coincidental but is related to exogenous factors.

While diabetes may result from dependence on imported food, there is much to suggest that the tumor diseases are a late consequence of radiation exposure from the nuclear bomb tests that were undertaken in the Marshall Islands. Another large portion of the homeless population are indigenous Hawaiians, disempowered and impoverished. White people make up only a small portion of the group. Although some even have jobs, their children still grow up in tents because they cannot pay the extremely high rents and security deposits. During 2024, several thousand students in the state of Hawaii were homeless at one time or another or living in conditions of unstable housing [75].

There is a great danger that this will become a permanent situation, because those who once live on the street have great difficulties in getting a permanent place of residence again. Left to their own devices and without proof of an address, bureaucratic hurdles are almost insurmountable. While tent settlements are growing on Hawaii's beaches, Facebook founder Mark Zuckerberg bought a *700-acre* plot of land on the Hawaiian island of Kauai. At the same time, he pursued lawsuits on hundreds of residents to prevent potential claims to small parcels within his property and to force a sale [76]. After the ensuing media disaster, Zuckerberg dropped the charges, but the land was eventually auctioned.

A similar development to Hawaii may be in store for Barbuda. Actor Robert De Niro is building a luxury resort called "Paradise Found" on the island. The national parliament has passed a new law, the Paradise Found Bill, which overrides the mechanism that gives the Barbuda Council decision-making power over projects on the island. Together with the Australian billionaire James Packer, De Niro thus obtained a 99-year lease concluded for a bargain price without the council's approval. Through the Paradise Found Act, which should probably be called "Paradise Lost," the investors are also granted a tax waiver for 25 years, which are ideal conditions for the exploitation of civilians and resources.

Besides the exclusion of the Barbudan people in the decision-making over the land and the economically adverse conditions, another issue is the destruction of pristine ecosystems. Barbudans even went to court in London to fight for their land (the country is part of the British Commonwealth) but lost their case.

After Hurricane Irma, Hollywood actor De Niro said he feels a responsibility to help and wants to create jobs as quickly as possible by building a hotel [77]. Immediately after the storm, he was in direct contact with the prime minister of Antigua and Barbuda, Gaston Browne. However, when asked about the people of Barbuda, De Niro admits that he hardly knows them. Some may accuse the Hollywood actor of a "White Savior Complex," meaning white people with supposedly positive intentions want to "help" Black, Indigenous, People of Color [BIPoC] without understanding the grievances they may be cementing [78].

Whether De Niro intends to do a good deed or simply wants to invest money profitably remains an open question. Even before the disaster, the prime minister appointed him special envoy for the country's economy. While negotiators from the national government of Antigua and Barbuda are making emotional speeches at international climate conferences demanding more climate justice for the benefit of the small island states, the country itself is circumventing laws and booting out its own population in order to leave the local environment to foreign investors. This is also part of the reality on the small island states, the interplay of light and shadow.

Toxic Algae Belt

One afternoon, my host takes me to a stretch of coast to show me one of the most beautiful corners of the island and, at the same time, draw my attention to another environmental problem. The sandy beach is littered with washed-up plastic garbage. Strikingly, a brown layer of algae floats like a blanket on the surf and extends several meters into the sea. Mountains of *Sargassum*, also known as gulfweed, also pile up on land. In the past 10 years, this algae species has increasingly appeared in the Caribbean, for example, on Barbuda and in Mexico, where it is already threatening tourism in some places. When *Sargassum* slowly decomposes on land, it produces a foul odor. Certainly, one would not like to swim between the meter-long brown loops.

Algal blooms can cause great harm in the ocean, as decaying dead algae release toxins and cause a lack of oxygen. This combination can even lead to the death of fish and other marine life [79]. Both the warming of the oceans and the input of nutrients into the water promote algae growth. In Brazil, industrialized agriculture in the Amazon region is introducing more and more fertilizers into the Atlantic, which eventually reach the Caribbean Sea via the ocean circulation, driving this development.

The problem has already reached threatening proportions—scientists found that a huge algae belt has formed from West Africa to the Gulf of Mexico [80]. With a length of 8850 kilometers, the enormous algae fields are threatening ever more beaches. In Mexico, millions of dollars are already being invested to clean up coastal strips or build barriers in the sea against the algae. Some startups and nongovernmental organizations are exploring to use the algae to produce soap, food additives, or even fertilizers, so that incentives are created to collect the pesky seaweed. In

Barbuda, such countermeasures can hardly be financed, and so tons of stinky algae regularly wash up.

Before I start my return journey, I ask a fisherman to take me to the shipping container that the storm threw hundreds of meters away from the island into the sea, like a toy. In a small boat, we sail past destroyed mangrove forests. Mangroves grow in shallow waters and protect the coast from storms because they break waves. But if the storm intensity increases, even the strongest mangrove trees can no longer withstand it. Here and there, a few green leaves and small bushes appear in between the gray twigs, slowly starting to grow again—2 years after the storm.

Now, we witness a rare natural spectacle: Frigate birds use the mangrove forests off Barbuda as a breeding and resting place. Numerous young birds and the occasional black male with a large red gular pouch cavort among the females. It is believed to be the largest frigate bird colony in the Western Hemisphere. Where they managed to hide during the storm remains their secret. They fly elegantly above our heads while the fisherman describes how he experienced cyclone Irma. The force of the hurricane was without comparison for the inhabitants. The damage it caused is far from being undone, even after years have passed. This is exemplified by the destroyed mangrove and reef systems, which translate into permanent loss of income for fishermen. It can take many years for the sensitive ecosystem to regenerate.

Grab-and-Go Capitalism

Despite the beginning of regeneration, the bird paradise is in danger. Another tourism facility, a huge golf course, is being built on the adjacent coastal strip, which, according to an initial environmental risk assessment, poses a threat to the entire wetland. The name of the project promotion raises doubts: "Peace, Love and Happiness." Behind Peace, Love and Happiness stands another billionaire, John Paul DeJoria, who, among other things, has made his money marketing Patrón Tequila, an overpriced spirit, and by founding Paul Mitchell, an overpriced hair product line.

While a part of humanity faces insurmountable borders and cannot overcome them even in the greatest need, there is at the same time a small class of people for whom the borders of nation states virtually do not exist. They can jet from place to place to invest, to party, and to make even more money. Whether these actions turn out to be harmful to people and the environment seems to be irrelevant. The stories of this grab-and-go capitalism of multinationally operating individuals and companies are sometimes so bizarre that one might think they come from a bad novel, along with all the self-chosen labels. They are, however, a part of our reality, the outgrowths of a globalization that turns some people into starving workers and catapults and others into space as billionaires. These lives, so different, are directly intertwined, which are consequences of our consumption that manifests injustices.

But in an economic system that has normalized the exploitation of people and natural resources, it is hard to find an exit strategy and durable solutions.

In her book *The Shock Doctrine*, globalization critic Naomi Klein coins the term disaster capitalism [81]. She describes the phenomenon, which can be observed again and again, that the special circumstances of catastrophes and the resulting chaos are used to realize neoliberal (economic) reforms, privatizations, land acquisition, and the like. Billionaires and their political henchmen are also trying to profit from the chaos left behind by Hurricane Irma on Barbuda—by converting the last original remaining spots on Earth into luxury resorts.

What gives hope is the resistance of many Barbuda residents who courageously stand up to the government and the promises of short-term profits. The fact that the community has not yet been broken by the destruction, trauma, and pressure of the national government is a testament to the resilience of the population and their bond with one another. But these human capacities are not unlimited. Increasing weather extremes demand direct assistance. To this end, many small island states in the Caribbean are calling for existing public debt to be forgiven if investments are made in local adaptation measures in their own currencies. Such debt relief could be an important step toward better disaster preparedness. However, it is uncertain whether debt relief can help an island like Barbuda, whose population is being betrayed by its own government.

Crises: Welcome to the Anthropocene

Barbuda, the Marshall Islands, Fiji, and Hawaii—places of humanity's longing for natural beauty—are acutely threatened. During my travels, I witnessed the richness and diversity of the islands and felt horror at what societies are willing to put at risk for the consumption of fossil fuels. Not for the benefit of human development, but for a lifestyle that takes everything and gives little. For the false promise: Buy yourself happy! For the consumption of animals, clothes, plastic, and places. For a lifestyle that serves permanent distraction and is paid for with a bounced check.

We have reached a critical point. The past years showed once again how much the risk landscape has already changed. The island nation of Vanuatu, the Solomon Islands, Tonga, and Fiji were hit by Tropical Cyclone Harold in the midst of the COVID-19 pandemic, and Tonga was also hit by the mega-eruption of a submarine volcano in January 2022. Thus, communities had to simultaneously fight the forces of nature and maintain infection control measures. These dystopian scenes illustrate that all future crises, whether economic collapses, geophysical disasters such as earthquakes, or health crises such as pandemics, will occur against the backdrop of a changing climate.

The increasing parallelism of crises is limiting our scope for action and pushing disaster management to its limits. This makes it all the more important for us to focus on prevention in the face of the many crises and to contain risks at an early stage. In times of climate change, this means reducing emissions and adapting, i.e., "avoiding the unmanageable and managing the unavoidable."

Eliminating the worst risks and bringing the remaining risks under control, such as the severe consequences for small island states under the Paris temperature pathway through adaptation finance, is a prerequisite for human development. If states fail to do so, the risk is to fall into a permanent crisis response mode. The first symptoms of this dramatic change are already defining life on the small island states.

Thesis

On the low-lying small island states, the struggle for survival in the climate crisis has already begun. The imminent demise of entire cultures is being accepted for the sake of the growth paradigm of fossil industries. The islands without a future are sending a stark warning signal—a last call to avert the global catastrophe.

Chapter 4
Conflicts Between Nomadic Herders and Sedentary Farmers in the Sahel

Europe's External Borders in the Desert ♦ Burkina Faso: Terror Under the Starry Sky ♦ Successful Farmers at the Limit ♦ Ethiopia: A Violent Peace Laureate and a Country in Transition ♦ Sahel at Crossroads

When I was given a chicken as a farewell gift, I was convinced that I would return to the drylands of Burkina Faso in the near future. Now, several years have passed, and the precarious security situation has not permitted another trip to Nouna in the northwestern part of the country, where I undertook my field research on climate migration 2017. At the time, the small West African country in the mighty Sahel region still had the reputation for being relatively politically stable. Burkina Faso, a country with diverse traditions that have developed in extreme climatic conditions, will be the focus of this chapter (Fig. 4.1).

First, however, I would like to look at the entire Sahel from a climate change and global security perspective. The effects of climate change in the dry regions of the Sahel create an additional opportunity for terrorists to pit ethnic groups against each other and fuel armed conflict. The combination of terrorist violence and climatic threat also poses an enormous challenge to international stability, but first and foremost to the affected countries.

Crises have taken foothold in multiple locations across the wider Sahel region over the past decade. Before the war broke out in 2020, I travelled to Tigray, Ethiopia, to ask people about the climate crisis. It is not only the conflict in Tigray that threatening the stability in Ethiopia; a huge dam project on the Upper Nile is also putting a strain on relations with neighboring Sudan and Egypt. This project touches the economic interests of a variety of countries: It is intended to meet energy shortages in Ethiopia, one of the most populous countries in Africa, but it may adversely affect the water needs of Sudan and Egypt and could negatively impact the natural environment.

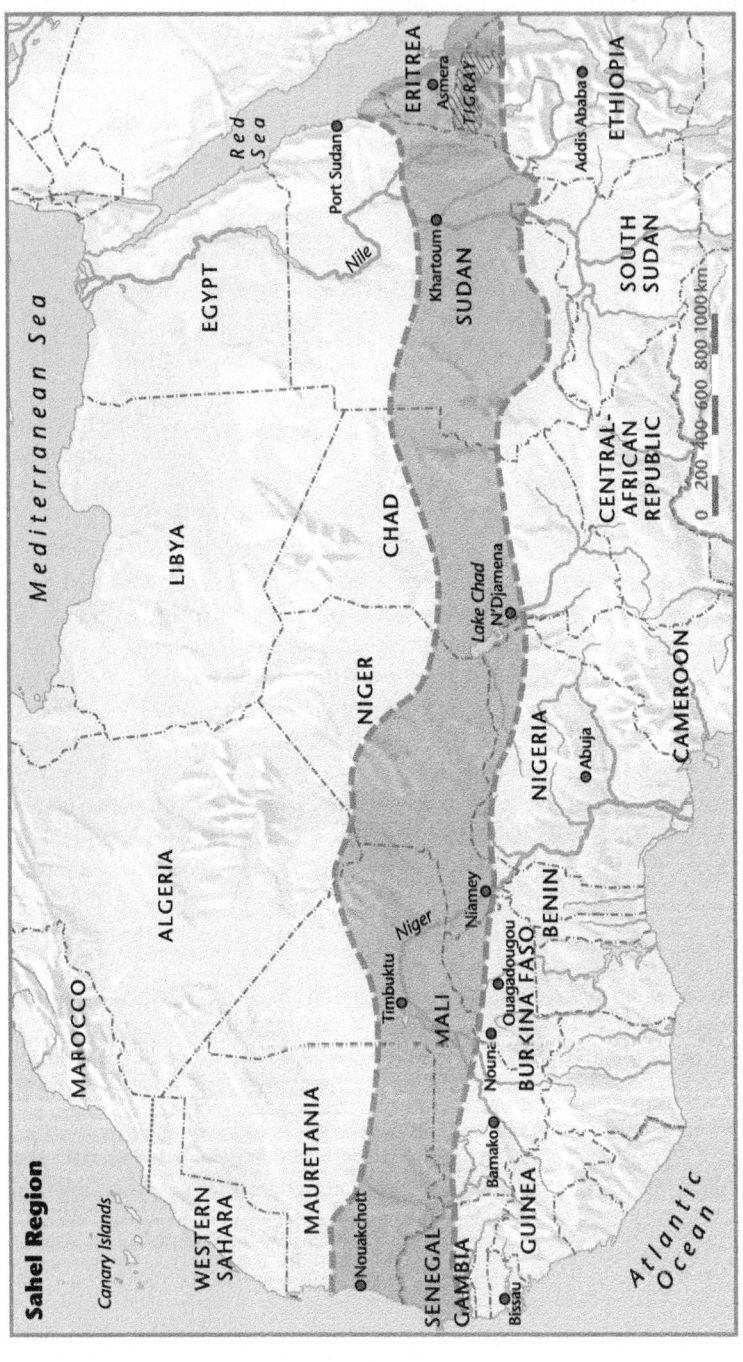

Fig. 4.1 The Sahel covers an area of more than three million square kilometers. There are different definitions of the zone, but in general, it includes parts of Senegal, Mauritania, Mali, Burkina Faso, Algeria, Niger, Nigeria, Cameroon, Central African Republic, Chad, Sudan, South Sudan, Eritrea, and Ethiopia

The Sahel comprises a vast area stretching from Senegal in the west to the northern tip of Ethiopia in the east [82]. This dry belt separates the Sahara Desert from the savannah landscape to the south. In recent years, the Sahel has often been in the headlines of global media. Mali's political and social destabilization as well as numerous terrorist attacks, such as on a hotel and a café in Burkina Faso's capital, Ouagadougou, in 2016, have had a lasting impact on the region. A series of military coups ensued, and Islamist groups, some of which show solidarity with the so-called Islamic State (IS) and Al-Qaeda, operate in large parts of the region. But governments are also participating in the violence. For example, the human rights organization Amnesty International reported that national military units carry out extrajudicial executions in rural areas in Niger, Burkina Faso, and Mali. Because of the poor security situation, many schools in these countries have had to close in recent years—the long-term consequences of this can hardly be assessed.

Under the umbrella of the UN peacekeeping mission MINUSMA in Mali (Mission Multidimensionnelle Intégrée des Nations Unies pour la Stabilisation au Mali), thousands of foreign soldiers, police officers, and civilian forces were operating in the Sahel from 2013 to 2023. Despite this concentrated effort, the security situation has been increasingly deteriorating. One reason is that Mali's vast territory can hardly effectively be brought under control. In 2021, a suicide attack was carried out north of the desert city of Gao, in which 12 German soldiers and a Belgian UN peacekeeper suffered injuries, some of them serious. The attack was attributed to the Al-Qaeda terrorist network. One goal of terrorist groups was to use Mali as a starting point to infiltrate other countries that had gradually improved their internal stability to weaken their governance. Where the state has no presence, lawless spaces emerge in the Sahel. Terrorist groups exploit this power vacuum, as do organized crime groups, which engage in drug trafficking and human smuggling, often with ties to the extremists.

Last but not least, our image of the Sahel is shaped by the inhumane migration that passes through the region via networks of smugglers. Often, it ends in death. Smuggling gangs abandon migrants in the middle of the desert and leave them to their fate. On the run as well as in the displacement camps, children and young people have hardly any access to school education. Women and children in particular are exposed to sexualized violence.

While much of the migration takes place within the African continent, some attempt to reach Europe via Libya, Algeria, Tunisia, and Morocco [83]. Ninety percent of all irregular migration to Europe occurs via the Mediterranean routes. The causes are diverse but bear certain similarities: people fleeing conflicts like the one in Mali; brutal oppression like the one in Eritrea; terrorist groups like Boko Haram in Nigeria; and hunger and extreme poverty like in Niger and large parts of the Sahel. Climate impacts, such as extreme droughts and heat waves, act as accelerators, amplifying these multifold drivers of migration.

Europe's External Borders in the Desert

The migration from a country in the Sahel to Europe can take months, sometimes even years, and involves extremely high risks. No one makes such a decision lightly—to exchange a familiar homeland for an uncertain future—even if many migrants are unaware of the extent of the danger involved. The disputes within the European Union over the distribution of refugees have led it to shift part of its border protection deep into the Sahel. In order to halt migration to Europe even before they reach the Mediterranean, cooperation with states such as Niger was intensified before the country itself experienced a coup in 2023. The result has been an increasing criminalization of migration that was previously tolerated.

The caravan town of Agadez in Niger, for example, was a hub for migration from Nigeria and other Sahel countries to Libya and on to Europe. Meanwhile, the national government was cracking down and drastically restricting movement on the routes from Agadez to the Sahara from 2016 to 2023 when the junta took over. The crackdown on migration drove smugglers to choose even more dangerous routes, with the result that more people were likely to die in the Sahara than in the Mediterranean. People crowd onto overloaded trucks which do not stop when a person falls off, partly because the vehicles threaten to get stuck in the sand if they hit the brakes. However, there are no exact figures for how many people lose their lives while on these migration routes—many families remain in the dark about the fate of their loved ones. The Missing Migrants Project of the International Organization for Migration (IOM) attempts to document where people perished during their migration—an almost impossible task. Researchers use the often unreliable data provided by national governments, interviews conducted with migrants who survived the trip, and reports from nongovernmental organizations and local media [84]. This painstaking and deeply depressing research helps families find closure and raises awareness that every single human life counts. However, the figures obtained through this detailed research are not reliable. The number of unreported cases is undoubtedly much higher.

The boundaries between migration, people smuggling, and the violence of extremist groups are becoming blurred in the public discussion about the situation in the Sahel. This also has to do with the reality on the ground. In practice, it is difficult to identify the division lines. Development aid and migration agreements are intended to suppress transnational movements, even though there are agreements of freedom of movement in place between individual states. Mobile border guard units in Niger, for example, were co-financed by German and Dutch taxpayers' money before the coup in Niger 2023. However, it is ultimately up to the national government to control whether these units act in accordance with human rights. The EU pledged development aid funds of more than 1 *billion* euros to Niger in the period from 2014 to 2021 [85]. These funds were used for social programs and as an impetus to create new jobs. Funding for adaptation measures to climate impacts also played a role. At the same time, Niger was to prevent irregular migration.

But there are other cross-border activities: A joint task force of five Sahel states (the so-called G5 Sahel: Burkina Faso, Niger, Chad, Mali, and Mauritania) was funded by the EU. The Bundeswehr was active in the Sahel states within the framework of the European Union Training Mission (EUTM) and the UN peacekeeping mission MINUSMA. It supported the training and education of Malian and the so-called joint force troops of the G5 Sahel states. In addition, there was a European Union Capacity Building Mission in Niger and Mali, where police officers are trained by the German police. The aim of German and European involvement was to combat terrorist groups in the region and stop irregular migration.

The second military coup within 9 months in Mali in May 2021, the military coup in Burkina Faso in January 2022, and the military coup in Niger in 2023 show how problematic German-supported military training can be in areas where democratic values do not apply or are not consolidated, and parts of the military commit human rights violations. Leading coup plotters in Mali had themselves participated in training programs in Germany. Russian mercenaries now support the military junta and fight for it against terrorist militias in Mali. Corrupt and violence-prone governments thus become part of an unholy alliance united in the fight against Islamist terror and, not least, against irregular migration toward Europe. The sustainability of such missions, in turn, also is called into question by the collapse of the Afghan government after the withdrawal of international troops in 2021. In light of the developments, France withdrew its troops from Mali in 2022. Germany stayed until 2023 and then also ended its decade-long engagement in the United Nations MINUSMA mission.

The interests of the European Union and the African Union are difficult to reconcile on the migration issue. Many African states are striving for greater freedom of movement, i.e., free movement of people and goods, as had been possible within the Economic Community of West African States (ECOWAS), for example. In the ECOWAS regional grouping, there was free movement of people and goods, but also the European Union pushed to introduce stricter controls to stave off migration to Europe. The EU practices free movement of people and goods, but wants to stop this in African states because of the threat of migration to Europe. Migration can be an engine for development and progress—an engine that is now stuttering.

The problems in the region are enormous in every respect. Poverty, corruption, and high population growth stand in the way of sustainable development. In Niger, for example, the population will have increased tenfold by 2025 as compared to 1950. According to the UN Department of Economic and Social Affairs, the population could increase sevenfold again by 2100—in a country that is half desert. Moreover, 12 of the 14 states that lie in or along the Sahel are among the least developed countries in the world. This is due not least to the extreme climatic conditions. Little rain and high temperatures require a high degree of adaptability. This is also how special forms of life and economy developed in the Sahel belt, such as the transhumance. Transhumance is a type of migratory pastoralism in which pastoralists travel long distances with their animals to ensure adequate supply of fodder by grazing different areas. They follow specific rainfall patterns and move across national boundaries. Transhumance was originally a finely balanced migration system that enabled people to raise livestock under adverse conditions. Sustainable management of nature was particularly important, so that in between periods of grazing of different livestock such as cattle, camels, and goats, nature could recover, and routes remained fertile grounds in the long term.

Over time, however, the centuries-old system has increasingly become a source of contention in the modern Sahel states. Violent clashes between nomadic pastoralists and sedentary farmers are becoming more frequent. The reasons are many: extreme poverty, changing land use practices, Islamist extremism, population growth, political divisions, and the heritage of the arbitrary drawing of borders by colonizers and the henceforth division of countries which disregarded common cultural and ethnic ties. Moreover, in recent years, shifts in rainfall have led to crop failures. Due to the unique geographic and climatic conditions in an arid and semi-arid region, respectively, such as the Sahel, land and water management is challenging and now further complicated by climate impacts. The deprivation of the livelihoods of farmers and pastoralists now poses a massive threat to the once peaceful coexistence of the people of the Sahel. At the same time, this development fosters the spread of terrorist networks and organized crime.

How climate change will affect the Sahel regionally is still characterized by uncertainties. This is partly due to the poor data situation, which is a problem in many African countries. Climate models rely on solid historical weather data. This means that datasets must be collected regularly from as dense a network of weather stations as possible. However, due to lack of funding, political upheaval, civil unrest, and war, most datasets in the Sahel are inadequate. Weather stations, some dating back to colonial times, have since been dismantled. Others provide data only irregularly. The number of weather stations per square kilometer is also significantly lower than that in the United States or the United Kingdom, for example, partly because large areas of the Sahel are very sparsely populated, and many governments do not see the need to invest scarce resources into new stations. Many countries have to focus their public finances on basic development projects, and very few would prefer the construction of a weather station to the building of an elementary school—understandably so. This, however, has dramatic consequences. Governments are unable to assess regional climate impacts with any degree of reliability in the medium and long term. Moreover, weather forecasting is compromised by the poor database. For city dwellers, this may be irrelevant or, at worst, a nuisance if they have not packed their umbrella or are inappropriately dressed. In agriculture, on the other hand, accurate data is crucial to a bountiful harvest. If the forecast is not correct, sowing, fertilizing, or harvesting may occur at the wrong time. Without modern irrigation, dependence on rainfall is high, and even minor shifts in rainfall patterns pose existential challenges to smallholder farmers. Because of these increasing difficulties, farmers and pastoralists are trying to diversify their incomes, sometimes practicing mixed forms of agriculture and pastoralism. However, when farmers themselves begin to raise livestock, the need for cooperation and symbiosis with pastoralists is noticeably reduced, which in turn fosters conflict.

If recurring droughts decimate livestock, the loss can translate into deep cuts into the social fabric. For nomadic groups such as the Fulani, Afar, and Tuareg, who are often linked by their religious affiliation or even by a common language, livestock is not only their central source of income, but also an important element of their cultural identity. As climate impacts affect sedentary farmers and nomadic pastoralists differently, tensions between these occupational groups can arise and also escalate. In addition, ethno-nationalist populism is used to mobilize and stir up resentment against "the others."

Hot Wars: Climate Change and Conflicts

The connection between climate impacts and conflicts is a hotly debated topic. More than 10 years ago, sociologist Harald Welzer published a book with the somewhat lurid title *Climate Wars: What People Will Be Killed for in the 21st Century* [86]. But do they really exist, the "climate wars"? Since the publication of the book, a lot has happened in research. Articles were published in renowned scientific journals showing statistical correlations between the occurrence of violent conflicts and extreme weather events, such as droughts [87, 88]. According to these findings, the risk of violent conflict can increase after an extreme weather event, especially in countries that are heavily dependent on agriculture and where there is ethnic polarization.

A 2019 study notes, "(...)Experts agree that climate has affected organized armed conflict within countries. However, other drivers, such as low socioeconomic development and low capabilities of the state, are judged to be substantially more influential, and the mechanisms of climate–conflict linkages remain a key uncertainty." [89]. That these other factors have thus far been considered more significant than the effects of climate change is not surprising. First, what can currently be observed are the early stages of the expected more drastic effects of climate change, and, second, the causes of violent conflict are multifaceted. Ultimately, a natural disaster can be the straw that breaks the camel's back.

One of the conflicts in which climate change may have acted as a threat multiplier is the war in Syria. A group of scientists pointed out that the Syrian civil war was preceded by the worst drought since temperature records began and contributed to the spiral of violence: Water scarcity, exacerbated by abysmal resource management, caused much of the country's livestock to die, prompting Syrian farmers to move to the cities in large numbers [90]. Resentment quickly grew over housing and labor shortages. At the same time, refugees arrived in Syria from Iraq to seek protection from the violence in their homeland. Food prices skyrocketed due to the drought. In parallel, there were demonstrations against the repressive and authoritarian Assad regime, inspired in part by the Arab Spring, which had a radiant effect on young people in particular. But the germinating hope for a more just future was bitterly disappointed. The bloody suppression of the demonstrations by Assad's henchmen provoked worldwide outrage and marked the beginning of the war against the Syrian people.

The history of the Syrian civil war is an example of the complexity of conflict genesis and of how many elements play a role in escalation. Would there have been a conflict without the drought? That cannot be said unequivocally. It may have been the straw that broke the camel's back, as described above. Either way, there would have been many opportunities for intervention in the emerging conflict, such as better social programs in urban areas, rural development, improved water management and adequate agricultural insurance for farmers and pastoralists, peaceful resolution of the protests and early concessions, to name just a few points.

How climate impacts affect the social fabric depends heavily on the particular infrastructural, institutional, and community capacities that are available. For example, if a storm hits a country with poor building materials, high political repression, and high socioeconomic inequality, the consequences are most likely to be far more severe than in a place with the best building standards, excellent disaster management, sound government, and functioning representation of minorities. First and foremost, climate protection, i.e., mitigation of emissions, is an important means of crisis prevention. Depending on where the weak points in the system lie, there are various intervention points to increase resilience, ranging from climate adaptation measures, such as better irrigation systems, to strengthening institutional capacities and classic development work, such as promoting education for women and girls, to environment-based mediation when a conflict of interest already exists.

Because of the variety of conflict drivers and because ultimately it is the individual, the government, or the nongovernmental group that decides whether he or she will use violence as a means to force their respective interests, the term climate war is at least fuzzy. However, no one can seriously question that climate change has an influence on violent conflicts. Where does the conflict start and which of its drivers are most relevant? Many researchers are likely to argue over an answer to these questions in the years to come, while climate change mercilessly runs its course.

In the Sahel, different drivers of conflict converge. Forced migration in response to the threat of violence can be both a consequence and a trigger of conflict. The latter is the case, for example, when competition for already scarce resources, such as water or land, intensifies and instruments for balancing interests are compromised. The relationship between dwindling livelihoods, migration, and conflict is convoluted and complex. The degradation of natural resources bears significant consequences for social systems. Especially if agricultural dependency is high and insurance schemes, which could prevent crop yield losses from turning into crises, are not prevalent. Even relatively stable countries can be shaken by extreme climate impacts, such as was the case in Burkina Faso.

Burkina Faso: Terror Under the Starry Sky

On the nearly 300-kilometer drive through Burkina Faso, large mango trees line part of the road we have to travel to reach the destination of our field research, the small town of Nouna, not far from the country's border with Mali. A colleague explains that many of the trees were planted during the French colonial period to provide shade along the long-distance roads. But a large part of the country has already been deforested, partly because arable farming and raising livestock require ever larger areas for the growing population, and lucrative cash crops, such as cotton, are replacing food crops like millet. Cotton plants need shade-free land, so trees are cut down—and thus a vicious cycle begins. Without trees, the water cycle changes; soils dry out, and yields decline. The rise in global mean temperatures leads to longer and more intense droughts. In many places, wood or charcoals are used as fuel for cooking. This, too, is detrimental to forest areas and also contributes to health-hazardous indoor air pollution. In this situation, people have few alternatives. After all, they depend on natural resources for survival.

I am traveling with a group led by Professor Rainer Sauerborn, who researches climate change and health at the University of Heidelberg. He took me to the region in the north of Burkina Faso where he worked as a doctor in the early 1980s. Since then, he has put endless energy into establishing numerous research collaborations at the interface of climate change and health in Nouna, especially with the "Centre de Recherche en Santé de Nouna." The health research center in the small town has established itself over the past decades and now reports to the Burkinabe Ministry of Health. Its research focuses on the spread of infectious diseases such as malaria and HIV in the region, but it also deals with the interface between climate impacts and human health, such as the effects of malnutrition.

When we arrive in Nouna, the destination of our trip, it is already late. We are advised not to move around freely in the village, as there have been isolated kidnappings of Western foreigners in the area. In order to increase our security, two young men, about 18 years old, with machine guns are posted near our huts. The armed teenagers do not necessarily add to my sense of security. But at this point, in 2017, none of us suspect that just 2 years later, this part of the country would be a no-go area. Burkina Faso, a state that long stood for stability and peaceful coexistence among different ethnic and religious groups, became a target for terrorist activity. Now, of some two million displaced people in the Sahel, about half are Burkinabe. How did this come about?

Extreme droughts, as well as uncertain land use rights, have contributed to a deep division between nomadic Fulani and sedentary Mossi. This divide, in turn, provides extremists with a breeding ground for sparking violence and recruiting fighters. Ansarul Islam, for example, a local terrorist group in solidarity with Al-Qaeda in the Maghreb (AQIM), uses ethno-nationalist propaganda to recruit young men from the Fulani group in particular. This strategy meets fertile ground because the situation of many pastoralists is precarious. Recurrent droughts have decimated the livestock of the Fulani, some of whom live nomadically, and the number of cattle per capita is falling, while at the same time the population is growing. Many herders are forced to sell their cattle; in the worst case, the animals perish in the desert sand.

The loss of livestock not only threatens livelihoods in economic terms, but also leads to uprooting and not seldomly into an identity crisis [91]. Traditions and social status are closely linked to cattle-herding practices. At the time of our visit, the ruling elite in Burkina Faso hardly includes nomadic Fulani; rather, the minority fills the prisons—often due to arbitrary arrests. The fact that there are virtually no judges, police officers, or military personnel among the Fulani further reinforces their marginalization. Ethnic discrimination and state corruption are commonplace. Because of the violence in the north of the country, many people are fleeing to the capital Ouagadougou, where they live with relatives or in temporary housing.

Successful Farmers at the Limit

It is not only nomadic pastoralists who are affected by extreme weather conditions; sedentary farmers are also struggling with climate change. In Bourasso, a village on the border with Mali, many people had already been forced to migrate before the outbreak of violence in the country because they had suffered hunger due to crop losses. The village was previously known for its successful farming. Crop surpluses of up to 20% were used, for example, to brew beer. The millet beer tastes slightly bitter and has a strong earthy smell. A Muslim trader jokes that he always drinks it under the table so that God does not see him. But everything is different in 2017. Repeating mini-droughts are destroying crops because there are no modern irrigation systems: "This year has been hard for us. Normally we work here in the area. Our land brought us a lot of fruits. But lately, there has been hardly any rain. In the past, if you did a good job, you could earn some money and you were content," reports 30-year-old Yacouba. The Chef du Village, a village mayor, explains to us with concern that hunger and malnutrition have forced a larger group of men from Bourasso to migrate for the first time [92]. It is mostly the younger men, fathers of small children, who decide to migrate. Hoping to feed their families with the money they acquire, they leave for a season, following the seemingly lucrative offers of recruiters who approach them in the villages. "We didn't hesitate and followed them there," Traoré recounts [93].

So the men went to Mali and the Ivory Coast to work as cheap laborers cutting sugarcane on industrial plantations or slaving away in gold mines. Occupational safety is nonexistent; medical treatment for injuries or malaria infections is sorely lacking. The wages stand in no relation to the physical torment. Gold mines have become an important source of finance for violent extremists in the Sahel and a hub for human trafficking. Gradually, it becomes clear in our interviews that the men were recruited under false pretenses. One reports, "The recruiters told us we would get one thousand francs (about €1.50) per meter of sugarcane. But when we arrived, there was no more talk of that." The migrants received neither the promised remuneration nor meals on site; they fell victim to fraudsters exploiting their desperate situation.

Moreover, 35-year-old Christian laments, "We grew crops here in the village, but earned nothing. That's why we left the village, hoping to get some [money] to bring back [...] When you go to the farm, you see that everything is barren. I knew if I didn't migrate during this dry season, my children and family would have nothing to eat. [...] I have nothing else to say. I have no cows, no sheep. I have nothing. I am now in the hands of God." Christian returned to his village of Bourasso after 3 months of working on sugarcane plantations in the Ivory Coast—with an injury to his hand from the sugarcane machete and the equivalent of 30 euros. Shame and resignation speak from his words. He tells us how much he misses his family. He is determined not to migrate again, come what may. After returning to the village, the results for the other men are also sobering. Many bring home only a small amount; some had their money taken away by their employers or had to pay large sums for their transportation. "Our boss disappeared with the money," several report. Only one villager actually earned well away from home. As a trained mechanic, he had better chances of finding a job in a city.

While the men tell us about the extremely stressful physical work, women tell us about the fears they go through. They ask themselves: Will their husbands come back? If so, in what condition? Many are also concerned that the returnees could bring back diseases, such as HIV. While the men work away from home, the women are forced to provide for themselves and their children; they must look for work in the area in order to survive. One woman whose husband has migrated tells us, "I work in other people's fields to earn money for millet. Sometimes I take a small loan to buy food. If we don't earn anything, we have to sleep with an empty stomach. (...) There are so many problems! Sometimes we don't even have soap to wash the clothes or money to have the grains milled." Often, the women are left to fend for themselves for months at a time. The work in the house and in the fields and the care work for the children weigh on their shoulders. The women are extremely critical of their husbands' seasonal work, and all too often, their hopes for a better life through additional income are disappointed.

Despite their traumatic experiences, many villagers see no other option than to try again and migrate once more when the harvests collapse again. This underlines the hopelessness of their situation and is at the same time a sign of their determined will to fight for the survival of their families. They have no alternative.

Overall, the situation of small farmers—and also nomadic herders—in Burkina Faso is increasingly deteriorating. In the absence of modern irrigation systems and new sources of income, migration, perhaps even permanent relocation, may increasingly be the last option left to them. Burkina Faso, like most other countries, does not have a comprehensive legal or policy framework for protecting or assisting people who migrate due to slow-onset climate events. The country has ratified the Kampala Convention (the 2009 African Union Convention for the Protection and

Assistance of Internally Displaced Persons in Africa), which recognizes that natural disasters can force migration. But all legal safeguards are of no use if practical implementation fails. Many courts in Burkina Faso are overburdened, and their authority weakened by corruption. Many Burkinabe are afraid of the police and the military—they cannot rely on the state. It is therefore all the more important that there are people who implement concrete changes and provide a glimmer of hope.

> **Modern Traditional Architecture**
> Francis Kéré has put his visions into action with innovative architecture in his hometown of Gando (Boulgou Province, in eastern Burkina Faso). In his buildings, beauty is combined with sustainability, for which he uses local materials as construction materials, for example, wood or clay reinforced with cement. The village community is involved in the planning process. This results in buildings that are adapted to the climatic conditions and have a pleasant indoor climate even when outside temperatures are high. This is an aspect of great importance, especially for schools, because buildings made of concrete and glass would quickly become an oven in such latitudes. Kéré's architecture promotes learning; his modern and at the same time traditional work inspires people, especially the youth.
>
> Based on an idea by the now deceased director Christoph Schlingensief, Kéré even designed an opera village in the savannah of Burkina Faso, 30 km northeast of the capital Ouagadougou. As part of the project, which has now been running for more than 10 years, boys and girls from the surrounding villages not only get something to eat; they also receive training: in classical school subjects, but also in acting, dancing, and singing. The idea of trying to change the world with music and architecture is coupled with realism. Learning and creating something to be proud of are important experiences while growing up. In the opera village, children and young people have the chance to build self-confidence and break out of the vicious circle of poverty.
>
> Francis Kéré came to Germany as a teenager on a scholarship. Out of a desire to give something back to his home village, which sent him out into the world, he began collecting donations while still a student in Berlin. When he graduated, he built a school in Gando, the first in his village. Today, Kéré teaches at the Technical University of Munich and passes his knowledge and vision on to the next generation of students. He has received several awards for his work, including the prestigious Pritzker Prize for Architecture in 2022. He uses his success to do even more for the people of Gando. With the Kéré Foundation, for example, new projects are being financed that are not just about buildings, but also about reforestation and local value chains. Given the massive problems in the country, such projects seem like a drop in the ocean. But the indomitable confidence associated with Kéré's work radiates hope. Using one's own skills and talents for concrete improvements and thereby creating impetus for greater change is at the core of human progress.

At the end of our eventful stay in Burkina Faso, new research questions are discussed, and German-Burkinabe author teams are formed. Learning from each other and from different academic disciplines is not a brittle theory here, but is at the heart of our exchange. In addition to my migration research, for example, I learn a lot about the spread of malaria and the tough fight against malnutrition.

For our last evening in Burkina Faso in Nouna, Rainer Sauerborn has come up with something special: We are arranging a musical exchange with the colleagues from the "Centre de Recherche en Santé de Nouna," who provide significant support for our research. The nationally known group Les Etoiles de Nouna is invited. They play on their balafons, a kind of xylophone that uses gourds (calabashes) as resonators and produces a wonderful sound. Rainer and I perform "Wochenend und Sonnenschein" by the Comedian Harmonists (he played the electric piano, I did the vocals); then, he plays another song by Debussy and invites the musicians to join in. Since it contains many triads, it fits perfectly with the pentatonically oriented balafons. This unusual mix sounds fantastic; the mood is exuberant. A praying mantis climbs over my plate. In the mango trees hang flying foxes as the sun is setting on Nouna.

The next morning, we set off, driving back to Ouagadougou over dusty roads. Our return to Nouna was already planned but could not be realized because of the escalating conflict. We stay with the beautiful and the distressing memories of the country.

Besides Burkina Faso, many other Sahel countries are struggling with climate change, as well as with socioeconomic challenges. The next subchapter will look at the other side of the continent, the Horn of Africa, where currently millions of refugees and internally displaced persons are staying. Almost a million refugees live in Ethiopia, which is itself affected by conflicts, droughts, and social division.

Ethiopia: A Country in Transition

What do you think of when you hear "Ethiopia"? The terrible droughts and famines of the 1980s that cost the lives of over 1.2 million Ethiopians? The UNESCO World Heritage Site in Lalibela, where churches carved out of rock inspire domestic and foreign pilgrims and tourists? Of President Abiy Ahmed, who was awarded the 2019 Nobel Peace Prize for his reconciliation with Eritrea, but soon after waged war in Tigray?

With its many facets, Ethiopia is a fascinating and complex country. Over 80 languages are spoken in the state, 5 of which are even official languages (Amharic, Oromo, Afar, Tigrinya, and Somali). The geography of the country is also characterized by a surprising diversity. Rock massifs seem to rise up from nowhere; the great African Rift Valley characterizes the landscape. Extremely fertile areas alternate with deforested wastelands. In many places, agriculture is practiced—two-thirds of the population work in this sector—and only slightly more than one-fifth of Ethiopians live in cities.

On my journey, which takes me from the capital Addis Ababa to Humbo in the south and Tigray in the north of the country, I talk to Ethiopian returnees who have had to migrate due to extreme drought. "Twenty-one days I walked on foot. From here to the sea. After 24 hours on a boat, we continued from Yemen to Saudi Arabia," Asefa tells us in a cactus grove in Tigray [94]. The 21 days were followed by 8 months of hard labor on Saudi construction sites. "They treated us like dogs - no, worse." Asefa had set out because a drought was severely affecting crops in the region. The landless farmer found himself deprived of his livelihood and all prospects for a better future. At the end of the 8 months, Asefa was deported by the Saudi authorities; he had entered the country illegally.

On the trip that I am making together with a group from the children's aid organization World Vision, we hear again and again depressing stories of people who, despite pressing problems, do not want to leave, who love their land and home. But more and more, this wonderful land is degraded by deforestation, overuse, and plastic waste. And the effects of climate change are becoming much more noticeable. We talk to farmers about hotter summers, unpredictable rainfall, and an uncertain future for them and their children. The question looms: What kind of home will Ethiopia have to offer its future generations?

In all emissions scenarios, the number of extremely hot days in Ethiopia will increase, from the current level of about 57 days per year to up to 74 by mid-century in a Paris consensus scenario and up to 151 by the end of the century in the high-emissions scenario [95]. The difference between a low- and high-emissions scenario is thus huge and could ultimately determine the future habitability of the country. Areas traditionally used by nomadic pastoralists would be particularly affected. Their economic situation would tend to worsen as a result. Since the mid-1980s, the rainfall that is so important for crop growth and characterizes the weather in East Africa from March to May has decreased. Rains often start later in the year and stop earlier. These changes pose enormous challenges because modern irrigation systems are scarce, and a delayed rain or brief drought can drastically decimate crops. How rainfall will develop in the long term in East Africa is still a matter of debate. Contrary to the current trend of increasing drought, it could become much wetter in East Africa in perspective. This paradox, for which climate scientists are still searching precise explanations, makes long-term adaptation planning difficult. Even increasing precipitation or changing rainfall patterns can lead to extremes, such as flooding, and in no way improve the situation of farmers.

One consequence of current environmental changes is migration from rural to urban areas, or from degraded areas to fertile regions. Or, as in the case of the farmer Asefa, even across national borders. As explained at the outset, migration decisions are usually made for several reasons. Both the threat of loss of income and social or political upheaval can prompt people to leave their place of origin. Climate change can alter these factors, and migration may be the last form of adaptation to environmental change. In Asefa's case, too, several things played together. He recounts, "There were no crops, no irrigation, no income from selling wood. There was drought here, often drought. Thus, I was forced to go to Saudi Arabia. I paid 18,000 birr to the traffickers (about 330 euros)." On the ground, he earned only about 100 euros per month, which he sent to his family.

In addition to climatic and economic constraints, a country's demographic development can also be a reason for migration. An Ethiopian woman still has more than four children on average. Many farmers we met had seven or eight children, who later have to divide the little land among themselves, unless they move to the cities or abroad. This demographic pressure also used to exist in now industrialized countries, like Germany. The emigration history of many Germans who moved to the United States in the mid-nineteenth to early twentieth centuries is also related to population growth in the countryside as well as natural disasters that destroyed crops. To be sure, the German story of emigration cannot be directly compared to the demographic trends in Ethiopia. But there are some parallels, for example, urbanization, which has already taken place in the industrialized world and is now in full swing in many African countries.

Addis Ababa, the capital of Ethiopia, is booming. Construction sites characterize the city, which is a place of arrival for many migrants from the surrounding areas. In many parts of Africa, urbanization is rapidly changing societies. A part of the rural population is increasingly drawn to the cities, where there are more opportunities for education and work. The destruction of agricultural livelihoods due to climate impacts and price dumping of agricultural products are also drivers of rural-urban migration. But the urban labor markets often do not have sufficient capacity to actually provide adequate work for all newcomers.

The potential of the growing young population is hardly used to advance society. Climate change with its multilayered consequences worsens the living situation of people not only in rural areas but also in cities, albeit in different forms. While in rural areas the population's agricultural livelihood is directly dependent on an intact environment, heat extremes in urban areas, for example, translate into reduced productivity and health risks. Water and food shortages can result in humanitarian emergencies, especially in rural areas, as well as trigger unrest in urban agglomerations through price increases of staple foods.

But the cactus grove near the border with Eritrea, where we conduct the interviews, shows a trace of hope for a better future. During our conversation, Asefa smiles again and again. We sit in the shade of a tree. Everything around us is green, despite a prolonged dry spell. After hours of driving through parched lands, we have landed in a valley where communities have joined hands to make the land fertile again using the Farmer Managed Natural Regeneration (FMNR) reforestation method. The method is as simple as it is ingenious. Instead of planting new trees, farmers use the existing, still intact root system for reforestation.

Tony Rinaudo, an Australian agronomist and worker for the children's charity World Vision, spent many years of his life in Niger and sub-Saharan Africa, where he gained experience in how natural reforestation can restore lost livelihoods. For this, he received the Right Livelihood Award in 2018. Movie director Volker Schlöndorff accompanied Rinaudo on several trips and produced the successful documentary film "The Forest Maker," which premiered in 2022.

Rinaudo has been passing on his knowledge for decades and can look back on more and more success stories of farmers whose lands have become fertile again. The central element of the method is cohesion within the village community, which is why FMNR also belongs to the toolbox of environmental peacebuilding [96]. In workshops, knowledge is imparted about the "forest under the Earth." Even in heavily degraded areas, there are still roots that can grow into bushes. With the right pruning, tender tree trunks can form and grow into shade trees within a few years. Under this shade, in turn, many plants develop especially well in particularly hot areas. In addition, the trees support the natural water cycle. In some areas, crop yields have doubled as a result of this method [97].

A prerequisite for the success of the projects is that the village community agrees not to cut down the trees prematurely, but to use only individual branches for firewood, fodder, or medicine. This also includes keeping livestock off the land initially so that goats and cattle do not graze over the young plants. In extremely poor areas, such as Niger or Ethiopia, reaching such a consensus to use the land only in a restricted way for a certain period of time is a diplomatic masterstroke. But in many villages in the Sahel, well-functioning social systems exist, and the community is tightly knit. People are looking for new ideas and impulses to give their children a better future.

In Mali, a pilot project was launched in which nomadic pastoralists and smallholders work together to promote reforestation. Traditional knowledge of local vegetation is crucial for success. For this reason, the village communities themselves have to identify the tree species that are to grow up on the basis of the roots [98]. Some trees are better suited later as a source of food for animals, others for obtaining traditional medicines or for raising forest bees. Every FMNR reforestation project is also a small experiment. A balance must be struck between how many trees are suitable for agroforestry and how their subsequent pruning should be done to ensure that enough sunlight reaches the ground. At the same time, development aid organizations such as World Vision are trying to reduce wood consumption through modernized cooking stoves. The communities bear the responsibility for the project. When the successes become visible, a multiplication effect often occurs: The farmers pass on their knowledge to neighboring communities.

The development progress made through FMNR cannot be denied. Stabilizing livelihoods helps people adapt to climate impacts in situ [99]. Forested areas can also buffer regional changes in temperature locally to a limited extent. Furthermore, additional income, for example, from the sale of honey or firewood, enables a special type of migration: for education. A man whose small farm is participating in an FMNR project proudly shows us pictures of his daughter, who was able to go to university in the nearby town. Reforestation has also had positive effects on crop yields for Asefa's home region and led to a rise in the water table. If global warming can be limited, the FMNR method can also offer people in other countries a chance to adapt locally to environmental changes and thus avoid having to move away. For Asefa, the decision has already been made. Pointing to the land behind him, he says emphatically, "I have returned and I am not leaving again."

In Tigray, we are given a jar of honey from one of the local development projects. Through reforestation, the ecosystem can regenerate, and new sources of income such as beekeeping open up. It is by far the best honey I have ever tasted. We talk about whether it can be exported as a delicacy. But the farmers reply that exporting it to the EU would require major investments in order to comply with food standards and make them verifiable. In such discussions, it quickly becomes clear that there are no easy solutions to break the spiral of poverty. Not to mention that outbreaks of violence are a setback for all prior efforts: Our trip took place in 2017. Three years later, Tigray is engulfed in war, the trail of hope in the cactus grove extinguished.

A Mighty Dam

In addition to the disastrous war in northern Ethiopia, there is a new area of tension and thus more reason for concern. One is related to issues of energy supply in a changing climate. The construction of the Grand Ethiopian Renaissance Dam (also known as "GERD") from 2011 to 2023, a massive Nile Valley dam designed to meet Ethiopia's growing energy needs, has been straining relations with neighboring Sudan and Egypt, dividing the region [100]. The dam is set to become the continent's largest hydroelectric facility. However, Ethiopia and the dam are upstream, and the massive barrier affects the water supply of millions of people downstream. Already, the filling of the dam has triggered serious diplomatic entanglements and disputes, even threats of war. In September 2024, Egypt issued a formal letter to the UN Security Council over the fifth phase of filling up the large water reservoirs. This is because if the dam is filled too quickly, it can lead to water shortages in neighboring countries. For Egypt, the Nile is a lifeline. Slow filling would be an option. But the Ethiopian government wants to put the hydropower plant into operation as quickly as possible. On the one hand, this has economic reasons, but GERD is also a national prestige object.

Through a type of crowdfunding, many Ethiopians contributed to the construction of the dam. The government issued government bonds for the purpose of financing, which were very popular. But the entire project seems oversized and is coupled with a fair amount of national pride, which has a chilling effect on neighbors and results in new tensions. At the same time, the Horn of Africa and a populous country like Ethiopia hardly have a future without large-scale energy projects. Since the dam is already in place, the question now is how to use it peacefully and as profitably as possible for everyone. Together with Belgian scientists, colleagues from the Potsdam Institute for Climate Impact Research (PIK) have shown in a study how the natural seasonal flow of the Nile could be imitated by additional wind and solar energy, thus also meeting the necessary water supply in Sudan and Egypt [101]. This coupling of water, wind, and solar energy could bring benefits to the entire region, as it would regulate the water supply in a sensible and environmentally sound manner. But important prerequisites for this are still missing: First, the conflicting parties need to return to the negotiating table to discuss the operational design of the dam; second, further investment is needed in renewable energy and in a functional power grid that includes neighboring countries and enables transnational power trading. Then, the dam could be a win-win not only for Ethiopians.

But so far, there is no solution in sight. If it is not possible to manage the mighty dam amicably in the medium term, the granaries of Egypt and Sudan will be put at risk. If agricultural livelihoods are significantly restricted or even destroyed as a result of the altered water cycle, the consequences are unforeseeable—from migration from the relevant areas to military conflicts.

Sahel at Crossroads

The colonial era still casts its shadow over the region, and at the same time, the climate catastrophe is making its way through the Sahara. The challenges in the Sahel are immense. However, migration from the Sahel also reveals the region's potential: a young generation which is determined to fight for a better future, who are seeking the ideals of the European Union, and leave everything behind for the hope of freedom and a self-determined life. Malian filmmaker Abou Bakar Sidibé documents this irrepressible struggle. In the documentary film "Those Who Jump," produced together with Moritz Siebert and Estephan Wagner, he shows the attempts of numerous refugees, mostly from the sub-Saharan region, to cross a wall from Morocco to Spain. No one who still feels their homeland can provide for them dares this migration, which puts their own life at risk—that much is clear after watching the film. Even if the images of degrading and life-threatening migration are temporarily displaced by other events, the crisis of global displacement remains in full swing.

What does the future hold for the Sahel? Will more and more refugees be housed in strictly monitored camps? Will the criminalization of migration continue? Or will the region's history take a very different turn? All of this is difficult to predict, but one thing is certain: Without resistance, climate impacts will destroy the livelihoods of more and more people and cause them to flee. Instead of fortified borders, livelihood opportunities are needed in the Sahel.

Thesis

The pressure on Europe's external borders will continue to increase, and European values will have to be measured by how much human suffering the EU is willing to accept for the sake of border protection.

Chapter 5
Superstorms: Long-Term Impacts in the Philippines and Bangladesh

Hurricanes, Cyclones, and Typhoons ♦ Climate Diplomats in Tears ♦ Unsafe Shelters ♦ New Climate: New Challenges ♦ Climate Crisis Hotspot Bangladesh ♦ Worse Extreme Events, Fewer Fatalities ♦ Megacity Dhaka in the Grip of the Forces of Nature ♦ Taking on Debt to Survive ♦ Tiger Conservation Versus Human Development ♦ Cities for Climate Migrants ♦ One-third of National Territory Under Water

Arthur, Bertha, Cristobal, Dolly, Edouard, Fay, Gonzalo, Hanna, Isaias, Josephine, Kyle, Laura, Marco, Nana, Omar, Paulette, Rene, Sally, Teddy, Vicky, and Wilfred. These are not the names of children in a school class, but a hurricane alphabet for the Atlantic Ocean that repeats every 6 years and is established by the World Meteorological Organization (WMO). Each storm area has its own list, as does the North Pacific or South China Sea, with locally used names. These are intended to be short and easy to understand so that effective communication can be made with the public when storm warnings are issued. Each tropical cyclone in a season is given a name according to the chronological order of its occurrence. The lists are reused. Only in individual cases, names are exchanged, when a particularly severe storm has claimed many lives or caused serious economic losses. In such cases, the name is retired in order to pay respect to the victims and also not to cause panic among people who still have memories of the previous storm events.

The last time the abovementioned hurricane alphabet was used was 2020. But the season did not end with Storm Wilfred. For the second time in history, the Greek alphabet was used for expansion, as a record of 30 hurricanes hit the Atlantic in 2020. The Greek alphabet then continued until Storm Iota, which caused tremendous damage in Central America in November. Never before had so many severe storms built up in the Atlantic during one season. And at the same time, this was in line with a multi-year trend. That is because the 2021 storm season was the sixth consecutive season in which higher hurricane activity was recorded in the Atlantic than in a normal year [102]. Incidentally, the Greek alphabet will no longer be used in the future because the storm names are too difficult to distinguish from each other in pronunciation. Instead, the WMO decided on a backup alphabet to accommodate future developments. Such hurricanes in the wake of climate change continue to drive migration. Although a return after a hurricane is possible in many cases, the very next storm, the next extreme weather event, often threatens to displace people again. Possessions and critical infrastructure frequently are not fully rebuilt in a timely manner. As a result, many people have no choice but to leave their ancestral homeland permanently. But where to?

Hurricanes, Cyclones, and Typhoons

What distinguishes cyclones from typhoons and hurricanes from cyclones? Basically, little; all three refer to tropical cyclones. Depending on where they occur, different labels are used. For example, in the Northwest Pacific, the storms are called typhoons, in the North Atlantic hurricanes, and in the Indian Ocean and the South Pacific tropical cyclones. These labels always describe cyclones with sustained wind speeds of more than 118 km/h. The storms form over the sea, a prerequisite being temperatures above 26.5 °C in the upper water layer, i.e., to a depth of 50 m [103]. If the storms hit land with high wind speeds, they almost always cause severe damage.

In its sixth Assessment Report, the Intergovernmental Panel on Climate Change (IPCC) emphasizes the increasingly disastrous consequences of tropical storms for nature and people. Rising sea levels increase storm surges and enlarge flood zones. At the same time, warming is increasing precipitation, so that wind and water combine to increase the scale of destruction. Tropical cyclones are generated by complex physical processes, making projections of their future development difficult. In addition, there are few data points for analysis because cyclones occur relatively infrequently, and satellite monitoring has only been possible for a few decades. Based on the existing database, the IPCC assumes that the proportion of extreme hurricanes has increased significantly [104]. Climate projections point in the same direction. Even if overall not necessarily more storms develop, the number of extreme hurricanes will increase. The wind speeds will also continue to rise, and with them the force of their destructive power.

In addition, there is evidence that in the wake of global warming, tropical cyclones are slowing down their pace, moving more slowly from point A to point B. This can have devastating consequences [105]. This slowdown can have devastating consequences, such as when storms spend longer time periods over populated areas. The resulting property and environmental damage also means that people may not be able to return to their homes after the storm subsides and the waters recede.

Fear-inducing images of the devastating effects of such storms reach us again and again. The nongovernmental organization Germanwatch regularly publishes a list of the countries most severely affected by climate impacts. On this climate risk index, six Asian countries are among the top ten for the period from 2000 to 2019 [106]. Tropical cyclones are among the most costly hazards of climate change. Especially where coastal areas have high population densities, the continent reveals its open flanks. Simple dwellings cannot withstand even moderately strong storms. The affected settlements then have to be evacuated, and people have to flee from the force of the winds. In the Philippines, a variety of unfavorable conditions—including severe poverty and ineffective warning systems—coincided with one of the worst storms in history.

Climate Diplomats in Tears

A few days before the international climate protection negotiations in Warsaw in 2013, everything changed in Tacloban, the capital of the eastern Philippine province of Leyte. History for the province is now sliced into two: the time before and the time after Typhoon "Haiyan" swept across the Philippines. In between were devastation, death, and chaos. For days, communication lines are disrupted; information from the worst-hit areas in the province hardly reaches the outside world. As the extent of the disaster gradually becomes clear, the Philippine delegate Yeb Sano gives a harrowing speech at the conference in Warsaw. He points to the destructiveness of a climate crisis that has been unleashed and calls for more emissions reductions as well as aid for particularly affected countries. Sano is fighting back tears because he has family in the disaster area. He says his brother is safe, but not everyone has reported back yet.

In later moments of silence, other delegates also struggle to compose themselves in the face of the harrowing images from Tacloban [107]. Ships block the roads. Houses have been razed to the ground, and trees have been blown down like matchsticks. Debris and rubble are on the streets. Until the end of the COP, the Philippine delegate goes on hunger strike. He is fasting for the climate and wants to increase the pressure on the negotiations. At the same time, his action is a sign of solidarity with the people in the disaster area, who have hardly any food and only limited access to drinking water.

In a way, Yeb Sano succeeds, as the "Warsaw Mechanism" for loss and damage is established at the end of the COP, a stepping stone for the loss and damage fund that is established almost a decade later. The Warsaw Mechanism calls for improved risk management and increased technical assistance. But even extensive foreign aid cannot compensate for the severe consequences of the hurricane. Some 4.3 million people had to leave their homes and seek shelter in evacuation centers or with friends and relatives. Dead bodies are still being recovered weeks later, and the total number of victims is more than 7000. The lives to be mourned, the trauma, and the irretrievable sense of security—these are all permanent losses.

Unsafe Shelters: Typhoons Destroy the Philippines

Typhoon Haiyan with its deadly storm surge caught people unprepared. Many could not understand the term "storm surge" and thus did not correctly understand the warnings. Due to the force of the storm and the rising sea level, the flood penetrated deep into the interior of the country and left a wide swath of destruction [108]. There was simply not enough time to evacuate. But even those who were still able to leave their homes were not automatically safe. Some shelters were located in the flood zone or were structurally unsuitable to withstand the raging forces of nature. Because of this, people in evacuation centers also lost their lives [109]. The survivors had no choice but to flee into the unknown.

The migration that Haiyan triggered was huge. Besides the city of Tacloban, a large number of smaller- and medium-sized cities on other islands were also affected by the disaster. Most of the evacuations were over short distances. Many sought shelter with relatives and acquaintances. Some sought out higher-lying regions and returned after the storm. Others were driven to rural areas and suburbs of regional centers, to Cebu, a city in the Central Visayas region, or to the capital Manila to escape supply shortages.

Since the Philippines consists of many islands, migration routes across provinces are difficult. Supplying relief after a disaster like Haiyan is a logistical challenge because virtually everything must be flown in. The Philippine government's disaster management drew much criticism. For example, several Filipinos died at a rice distribution point when thousands of desperate people stormed the building, causing a wall to collapse. Although security forces were present, they did not manage to bring the situation under control quickly enough.

In addition to national disaster management, a large number of foreign aid organizations are usually active in the affected areas after such tragedies. In crisis situations, nongovernmental organizations such as ShelterBox try to provide displaced people who have lost everything with the most basic necessities [110]. In so-called survival boxes, they assemble blankets, tents, mosquito nets, cooking utensils, and small stoves to provide practical help on site. Tools are also included so that those hit so hard can build emergency shelters and repair their homes themselves.

In the meantime, the city of Tacloban has been rebuilt. Financial and technical aid from all over the world, as well as the tireless efforts of the residents, have made the reconstruction possible. The slums that were located in the flood zone are also back in place. Many slum dwellers have no other choice—outside the city, they may lose their source of income from fishing or selling goods. Or they have to spend additional funds to commute, resources which they simply do not have.

But what happens when the next disaster strikes the region? According to a government plan, more than 205,000 households are to be resettled so that they are not exposed to any new dangers from flooding in the long term [111]. But what a humane resettlement might look like and where those affected can move to is still largely unclear. Although some replacement housing has been built, the settlements offer such poor living conditions that those housed there refuse to use them permanently. Out of helplessness, the city of Tacloban designated restricted areas, declaring previously inhabited areas as "no-build zones." But fishermen and workers with other local activities had no alternative but to return there [112].

> **How Climate Change Gives Rise to Human Trafficking**
> In the aftermath of extreme weather events, criminal groups can take advantage of chaotic situations. In particular, women and children become targets of human trafficking or forced prostitution, but men also fall into dependency and have to perform forced labor, on ships or in quarries or agriculture. Sectors that are often hardly subject to control, such as fishing, are particularly affected. The loss of social structures, the disruption of communal ties, and poverty drive many people into desperate situations.
>
> Also, during involuntary migration, people take risks they would avoid under normal circumstances, but in order to survive, the displaced must come to trust complete strangers—among them criminals who shamelessly profit from their plight. Many have no bank accounts, no savings to use in times of need. When disaster strikes, they look for solutions to get their families through. When resources become scarce, parents also take their children out of school to work. Or a child may be forcibly married or promised to someone in order to secure the lives of the other children in the family. Girls are often affected by this. But boys also fall victim to human trafficking. They are lured from their parents' care by criminal gangs with false pretenses or with the promise that they can start an education elsewhere and lead a better life. Particularly dangerous are situations in which children and young people are left to fend for themselves, for example, if they have been separated from their parents or have been orphaned by disaster.
>
> Human trafficking was rampant after Typhoon Haiyan [113]. In the region where Haiyan unleashed its disaster, human trafficking for the purpose of prostitution and moonlighting under inhumane conditions existed even before, but the situation worsened as a result of the disaster. Women from the provinces were lured to big cities ostensibly to work as waitresses, but then forced into sex work [114]. Children were abused as cheap labor. People were even trafficked across national borders, for example, to other Asian countries or the Middle East [115].
>
> International organizations, such as the International Labor Organization (ILO) and the International Organization for Migration (IOM), as well as local aid organizations, are trying to mitigate the negative collateral effects of extreme weather events, for example, by helping vulnerable people to enroll in government aid programs and by providing education. In the case of

Haiyan, for example, cash-for-work projects were set up in which people could participate in reconstruction and other infrastructure measures and receive direct payments in return.

As extreme weather events increase, severing the social fabric, criminals will devise new ways to expand their reach. Scientists found that the loss of wildlife populations also increases the need for cheap labor, thus encouraging organized crime [116]. This makes it all the more important for organizations to advocate for the protection of vulnerable people and to reduce environmental risks. Those who have fallen into the clutches of traffickers but managed to escape them have a special role to play. They are in a position to provide guidance on how trafficked persons can be helped without putting them at further risk. Equally central are organizations that give people who were formerly forced into prostitution the chance to earn an income, because even if they were able to free themselves from enslavement, they continue to suffer from social stigmatization.

New Climate, New Challenges

The storms hitting the Philippines are becoming increasingly violent. In December 2021, Typhoon Rai, a Category 5 hurricane, caused damage in seven provinces of the island republic. In total, about 622,000 people were displaced. But hurricanes are not the only threat to the livelihoods in the Philippines. A whole range of other climate impacts are affecting the island nation and its inhabitants, increasing their vulnerability to extreme weather events in the long term. In 2020 alone, there were more than 4.4 million new internally displaced persons due to natural disasters. These included typhoons, flooding and also non-weather related volcanic eruptions. Gradual changes, such as droughts, can also alter migration patterns.

For example, the droughts that occurred in the very fertile Lake Sebu region as a result of the 2015 El Niño event (see Chap. 6) caused agricultural losses of more than $300 million. Families split up to earn a living across different places [117].

The consequences of climate change are increasingly becoming a factor in this chain of disasters, migration, and displacement. This is also reflected in rural-urban migration. The most affected fishermen and farmers are driven to the big cities, where they try to build up a new existence. This internal migration contributes to the continued growth of cities such as Manila. In contrast to the exceptionally beautiful rural regions of the Philippines, the capital Manila is developing into a juggernaut that continues to expand. The contrast between the shiny facades of the business districts and the informal settlements is stark.

The uncontrolled growth has led to inner city traffic in which even short distances can take several hours of travel time. During a stay in Manila, I try to get around on foot one late afternoon. I squeeze through lines of cars, past honking mopeds and jeepneys, and the garishly colored minibuses that originated from converted American military jeeps and have recently been replaced by safer and more environmentally friendly minibuses. But many small business owners of jeepneys cannot afford to buy the new vehicles, and their customers cannot pay higher prices for the rides [118]. Although the city center is already suffocating in the traffic chaos, decision-makers tell me that they seek to fulfil the population's desire for more private cars. This sounds crazy, but it follows an inner logic. After all, the worse the traffic is, the greater the impetus for the individual not to move around unprotected as a pedestrian or by bicycle. Public transportation is currently not an attractive alternative either: It is hopelessly overloaded.

Today, about 14.4 million people live in Manila. Without intervention, megacities will continue to grow, partly because the livelihoods of people in the hinterland are being destroyed by climate change impacts. In addition to hurricanes, there is frequent flooding in coastal areas, and heat waves make life difficult, especially for those who work outdoors. And the future looks even bleaker. The new normal is a much more severe risk landscape. For example, the Philippine Climate Change Commission estimates that 13.6 million residents in the Philippines may have to be relocated due to sea level rise alone.

The question is: How should such a challenge be met? Or will people be left to their fate, like the millions who already live in slums, in some cases even on the graves of cemeteries? This depressing prospect calls for more capacity building on climate change and migration [119]. On a visit to IOM in Manila, I inquire how and what the UN agency is working on in the Philippines. The country, which has seen a wide variety of migratory movements, has—in percentage terms—a diaspora that is among the largest in the world: More than 5% of all Filipinos live abroad to earn a living for their families and dependents. The IOM advises the Philippine government on creating effective structures for labor migration and enacting appropriate legislation. It also reaches out directly to trafficked persons and educates them about their rights. This is intended to protect potential migrant workers from exploitation, enable them to assert themselves abroad, and give them a chance to get better jobs. In this sense, international labor migration can be a form of adaptation to worsening environmental conditions—even if the hurdles for particularly affected people are generally higher for transboundary migration than for internal migration. If short-term evacuations are necessary after extreme weather events, the UN organization also provides disaster relief, including temporary shelter and psychological support.

In addition to the very concrete, practical work of aid organizations and UN institutions, research on regional climate impacts and weather forecasts is central to effective long-term disaster management. One important research institute dealing with climate impacts, air pollution, and sustainable development is the Manila Observatory, which was founded by Jesuits in the nineteenth century. When I visited it in 2015, I was impressed by the passion of the scientists on site. Under sometimes adverse conditions, they carry out practically relevant work, the results of which are often included in the IPCC reports, in which the state of climate research is systematically recorded. The Manila Observatory team is united by a common mission: to help the people of the Philippines through science. The institute, which also conducts research on tropical cyclones, sent aid workers to the disaster area after the devastating Haiyan storm. In order to counter the effects of climate change in countries like the Philippines, regional research institutes like this deserve far greater support than they have received to date.

Climate Crisis Hotspot Bangladesh

Like the Philippines, the South Asian country Bangladesh is facing complex climate impacts. With storms, river erosion, crop failures, flooding, and salinization of soil and drinking water, Bangladesh has become a climate change hotspot. Located in a river delta, the country is inhabited by 163 million people. The densely populated south of the country has to bear the brunt of increasingly frequent extreme tropical cyclones. Cyclones Sidr (2007) and Aila (2009) hit southern Bangladesh in quick succession. Two extreme storms, Fani and Amphan, also hit the delta in 2019 and 2020. The rural population has little capacity to adapt to the forces of nature.

> **Weather Extremes Affecting Refugee Camps**
> The strength of a storm is not the sole determinant of the extent of damage it causes. The losses depend on the number of people living in the affected areas, the infrastructure, and the physical assets. The more precarious the housing conditions, the less protection against wind and floods. In cases where the economic costs of destroyed poverty-stricken settlements are low, the human costs rise all the higher. Families who barely own anything lose everything. And they have no social safety net to catch them. Cyclone Mora, for example, had "only" wind speeds of up to 150 km/h. This meant it was considered weak (Category 1) according to the Saffir-Simpson scale, which categorizes hurricanes according to their strength. Cyclone Fani and Cyclone Amphan were classified as Category 5, with winds of up to 270 km/h. However, Mora also brought immense human suffering because the storm hit refugee shelters with all its force.
>
> Bangladesh took in some 900,000 Rohingya refugees, a persecuted minority from Myanmar. Storm Mora, which swept across the southeast of the country, dismantled the refugees' tent cities and shanty towns that had formed around the town of Cox's Bazar, like the Kutupalong Camp. Thousands of people were once again left with nothing. The few food supplies and laboriously built emergency housing were razed to the ground by gusts of wind and masses of rain. But despite the camp's exposure to extreme weather events, the number of residents continued to rise after the storm. Thus, the risk of violence and persecution is traded for the risk of the forces of nature [120]. Today, Kutupalong is the world's largest refugee camp, with over 600,000 people [121]. Many of them have been living in the limbo of the camp structures for several years.

> In turn, refugee movements and resettlements led to environmental damage in Bangladesh. Dense forests around Cox's Bazar were cut down by refugees in their need for firewood for cooking and building huts. Around 40% of the forest has thus disappeared since the 1990s—a vicious circle, because degraded areas are even more vulnerable to extreme weather events than intact ones. Without the tree cover, there is nothing to prevent landslides; the soil absorbs less water, and heat builds up in densely populated settlements. When major floods occur, communicable diseases increase. The UN Refugee Agency is trying to counteract this fatal development by providing gas stoves for households so that no more firewood is needed and by taking reforestation measures [122]. Fast-growing bamboo is to be used as a building material in the future.

Worse Extreme Events, Fewer Fatalities

Overall, Bangladesh can actually look back on a positive record in disaster management. The number of fatalities from tropical cyclones has fallen sharply compared to the 1970s. Infrastructure measures such as shelters, better early warning systems, and education about the risks posed by tidal waves have strengthened resilience. The fact that the Amphan supercyclone was approaching, for example, was recognized early on. Warnings and evacuations followed. Rapid information sharing through social media such as Twitter (now "X") and messenger services helped to quickly inform residents in different areas [123]. About 800,000 people left their homes as a precaution, and a total of 2.4 million people were internally displaced. About 100 people died, many from fallen power poles that caused fatal electric shocks in the flooded areas. Each fatality is one too many. Still, the casualty figures are markedly different from those of previous years, although an exact comparison is methodologically difficult. In 1999, when a similarly severe cyclone struck Odisha and the Bay of Bengal, it claimed 15,000 lives. In 1970, Cyclone Bhola was estimated to have killed half a million people in Bangladesh (then East Pakistan) and West Bengal (India).

Efficient evacuation plans, construction of sea walls, and government programs for disaster preparedness in conjunction with international development aid helped to ensure that today's storm surges meet barriers. At the same time, any adaptation measure is a race against time, as climate impacts become increasingly severe in the delta state. Maintaining functional warning systems over long periods of time is costly and requires the trust of the population in the responsible authorities. This is not always the case when corruption rates are high. Moreover, if the highest warning level is called out too often without anything serious happening, people tend to start taking it less seriously. It is therefore necessary to weigh carefully when warnings should be given so that those affected can react accordingly.

Protecting human lives from the direct effects of disasters through better forecasting is one thing. Rebuilding destroyed infrastructure is yet another. Economic damages carry long-term implications for the development of regions, and the setbacks are often felt by the population even years after the disaster occurred. People cannot return when the soils are salinized and their houses are still in ruins. Thus, some internal displacement, forced by tropical cyclones, also can lead to a permanent change of location, as I found out in my field research in Bangladesh.

Megacity Dhaka in the Grip of the Forces of Nature

My field research in Bangladesh took place in 2014 and served to gather information for a development cooperation project that would later provide trainings and support to people displaced by climate change. The objective is to explore what people's needs are who move from rural areas to the surrounding cities due to climate extremes. Two students, Rashed and Naoshin from Bangladesh, accompany me and help with the translation.

Our research trip is off to an inauspicious start. After I arrive in Dhaka, a travel ban is imposed. We are not allowed to leave the capital because of the threat of terrorism. The security situation is too precarious, they say. Bangladesh is repeatedly targeted by Islamist extremists; just 1 year later, the Italian aid worker Cesare Tavella is killed while jogging near my hotel at the time. In 2016, a total of 29 people died in an attack on a popular café in Dhaka.

We use the extra time for a visit to an informal settlement, where we conduct a first round of interviews, trying out the comprehensibility of the questionnaires we prepared. The Korail slum is the largest slum in Dhaka with an estimated 250,000 inhabitants. Exact figures are not available, many people are not even registered, and the boundaries between the slum and the rest of the city are fluid. In addition, there are many so-called pavement dwellers, i.e., people who live on the streets without any shelter at all. Dhaka has about nine million inhabitants; when the suburbs are included, the number roughly doubles.

Climate Impacts in Slums

Globally, almost one in eight people lives in a slum or informal settlement. The number of people living in these abysmal conditions already exceeds one billion. By the middle of the twenty-first century, it could increase by another one to two billion people due to urbanization and population growth. Climate impacts affect slum-dwellers particularly hard. The heat under the corrugated iron roofs, the overcrowded spaces, the lack of proper sanitation, and open fire cooking stoves put extreme burdens on the inhabitants.

Many slums are places of arrival for climate migrants who come from rural areas and try to get by with jobs in the informal economy, working in construction or factories or as housemaids. In the process, some become part of global supply chains, producing clothes, cheap decorations, or electrical appliances for European or American customers. With their minimally paid labor, companies maximize their profits.

Figure 5.1 illustrates that the world's largest slums are exposed to a whole range of climate impacts. If climate change impacts continue to intensify, the urban poor will be disproportionately affected because they typically live in exposed areas, such as hillsides or floodplains, and lack the financial resources to increase their adaptive capacity. However, although site-specific analyses would be urgently needed, studies are mostly limited to regional climate impact assessments. The Sustainable Development Goals (SDGs), agreed to by the international community in 2015, demonstrate that there is a global consensus on the urgent need for sustainable solutions. However, progress on implementing solutions has been painfully slow for the urban poor.

In the Korail slum in Bangladesh, we meet two women aged between 20 and 30 who moved to Dhaka from the south of the country just a few months earlier. Despite the extreme poverty in which they live, the women radiate dignity. One wears a colorful sari—in my view one of the most beautiful pieces of clothing invented by mankind—and the other a salwar kameez, a combination of wide pants and a long tunic. Moyna came to Dhaka to find work. Hosne never thought she would live outside her village. But then the mighty Brahmaputra swept away her land, and she fled to the capital with her family to survive. Moyna also describes how her old home, a town near the eastern border with India, was repeatedly flooded by heavy rains. But also the Korail slum is affected by extreme weather, which we soon get to feel.

Hosne says that they can barely make ends meet financially. Her husband borrows a bicycle rickshaw every day for about 10 euros and offers himself to passersby as a driver. He has to work several hours just to earn the rental fee. With luck, he brings home a small profit. When it rains, the customers usually stay away; then, his earning potential is close to zero. The family pays rent for their shack, but could be evicted from it any day, because the landlords, local gangsters, have no official land rights to the land they lease. In the dark, tiny dwelling where Hosne and her husband live with their three children, there is a cot on which they take turns sleeping. Everyone who can not find room on it sleeps on the floor or during other periods of the day or night. Five large families share two latrines. The women show me wounds on their lower legs—rat bites that have become infected. When a heavy monsoon rain occurs, everything is under water, making it difficult for the wounds to heal. The children also suffer greatly from the rat bites at night.

When we visit, it is dry season. But unexpectedly, heavy rain starts to fall. A sign of climate change? That is hard to say, as such oddities do occur—in recent years, however, more and more frequently, as the women tell us. They give us shelter and offer us to sit on the cot. The rain on the corrugated iron produces a roaring noise. Within minutes, the water rises dangerously high and slowly spills into the shack.

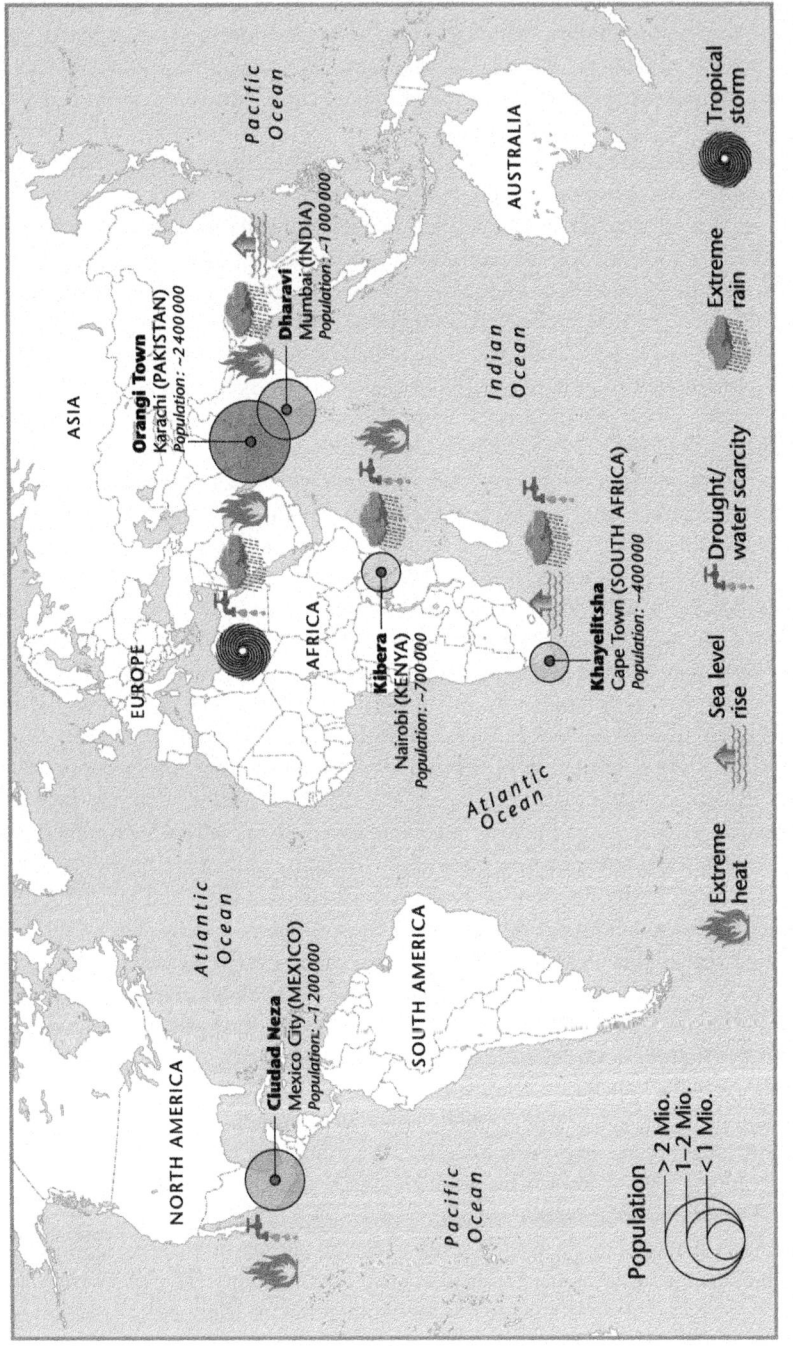

Fig. 5.1 Climate impacts in the largest urban informal settlements, schematic diagram

"That's how it is here, there's no real drainage system," Moyna says. We wait for another 20 minutes for the rain to let up a bit. Then, we wade for half an hour through the flood, where feces and dirt mix. Life expectancy in slums is low and infant mortality high. Health risks determine people's lives from birth.

At a street corner, we take a cab. Even the paved roads are now partially under water. "Welcome to Bangladesh, here you will experience a nightmare in broad daylight," says the driver only half-jokingly and ponders whether this could be a good saying to attract tourists. I am dropped off at the hotel, where I go to my comfortable room, stuff newspaper in the insides of my shoes for them to dry, and get into the shower. I think of Hosne and Moyna, who have no access to a bed and basic sanitation. A few kilometers away, they sleep in their huts, as they do every night.

Shortly after, the travel ban is lifted, and we can set off. On the drive from Dhaka to the south of Bangladesh, it becomes apparent that almost every square meter is used for settlements or agriculture. So where do migrants go when they move away from storms and sea level rise?

To find this out, we interview local experts, municipal officials, and also those who are always well informed about the latest developments in their city: tea stall owners. With their help, we identify new settlements in the slums where displaced people have recently settled. There, people describe to us what prompted them to migrate: "First we lost our agricultural land, and after three years our homestead land. Then we received land from the government. But we lost that too after two years," says a woman in Barisal. It is a typical story of displacement due to environmental change [124]. Meanwhile, this woman's husband, a former farmer, borrows a handcart every day to transport all sorts of things and feed his family with the little money he earns. Others work in rice mills or brick factories. But his income varies widely. "It's just enough to survive here. We were better off in the village, where we could farm our own land. Now we live like beggars," the woman complains.

Due to their dependence on fertile land and because they have no access to education, subsistence farmers have neither the financial resources nor the necessary skills to build a new livelihood in urban centers. About 43% of the population is engaged in agriculture. Their possessions mostly consist of a small plot of land, livestock, and shelter. Since they only generate a small income, a loss of these resources, for example, after a flood, can force them to relocate without any savings.

In another city in the south, I meet a family who recently had to relocate from the rural area and took their only cow with them to the outskirts of the city. The family lives and sleeps next to the animal, separated only by a thin plastic tarp. The odor of the animal dominates the room. During the day, a family member leads the cow by a rope to a nearby grassy area to graze. The animal is everything the family owns and hence is guarded and protected.

In Khulna, a city of around one million people near the Sundarban forests in southern Bangladesh, we meet a former fisherman in one of the sprawling slums. Abir is 30 years old and trying to make a living with a borrowed cargo rickshaw. The hard physical work has worn him out; he looks easily 10 or 20 years older. Storm Aila destroyed his small hut and the surrounding land he farmed. Everything was swept away by the floods. As a result, he moved toward Khulna, and for the first few months, he lived somewhere along a riverbank. Then, he found shelter in a slum. Because he could not pay the rent, he moved to the next slum. That is where we meet him. The heat builds up under the corrugated iron roof. Abir tells us about the time before the big storm, before Aila. At that time he earned well from fishing, he reports. But today, he says, there are hardly any fish left in the rivers. "I have no choice. Where would I go? I would go back to my home village if my house was still there." His two brothers still live there. Both continue to work as fishermen, but they no longer have a home; they live on the streets. Abir's fate represents countless similar fates in Bangladesh, people whom climate change has turned into nomads, uprooted without a foothold. They persevere in the places of arrival—until the next stroke of fate, the next storm, or hunger drives them elsewhere.

Because climate change affects so many, the local authorities are overwhelmed. When natural disasters occur, people first flee the countryside for the nearest cities. But even urban areas are not immune to severe weather. Floods and storm damages accumulate, and chaos regularly reigns after a disaster: How many people have flocked to the urban area? Where are they finding shelter? How do they feed themselves? No one knows the answer. Data on the influx of displaced persons is scarce. Ultimately, the nomads are left to fend for themselves and are not infrequently at the mercy of crooks and organized crime.

"There is no plan," the employee of a city administration freely admits when I ask him about the preparations for the next storm. The tone in his voice reflects that this is not due to a lack of will on the part of the officials, but rather to the simply overwhelming multitude of problems that the administration is far too small to handle. These include establishing a functioning sewage system, garbage collection, and the enforcement of minimum safety standards for new buildings. Even the most basic needs are not met due to lack of finance and capacity.

Taking on Debt to Survive

If IDPs have no family networks in the city, this means that they have to take on debt to survive. State aid is scarce, and the living costs in slums—paradoxical as it may sound—are expensive [125]. Even for extremely inadequate and insecure housing, sometimes horrendous rents have to be paid. Families who do not have their own toilet often pay for access to sanitary facilities. Due to poor sanitary conditions, diseases are rampant, which in turn can lead to health costs and loss of earnings. If there is no drinking water, it must be purchased as well. Because they earn so little, many slum dwellers cannot afford to buy products in large quantities. Or they do not have a refrigerator to store food for more than a day. This means they often have to resort to micro-packaging, which is usually produced by international companies such as Nestlé or Unilever and is more expensive overall than purchasing larger containers. Unprocessed, fresh food is increasingly giving way to packaged products from international companies in urban slums. The cost of food and water may even increase in the future due to more severe climate impacts such as extreme droughts. Despite differences in the cost of living, the line for extreme poverty as it is calculated by the World Bank does not distinguish between urban and rural areas and stands at only $2.15 a day. Thus, many people living in extreme urban poverty do not fall below the lower poverty line of the statistics.

The typical activities of migrants can also lead into a debt trap. For example, workers reported that they cannot work in brick factories and rice mills during the monsoon season because the products have to dry outdoors. At the same time, however, they live on the factory premises and become indebted for unpaid rent to the owners, who in some cases even confiscate their personal documents as security—a clear case of debt bondage. In Bangladesh, it is clear that the migrants' self-determination is further compromised by the effects of climate change. In the context of climate change, migrants are sometimes portrayed as free agents seeking a better life. Such a portrayal, however, is deceptive. Many of these people have lost any leeway to make decisions about their future.

Tiger Conservation Versus Human Development

Several interview partners tell me that after the cyclones, tigers occasionally invade villages and attack people. The tigers no longer find prey in the Sundarbans, the mangrove forests in southern Bangladesh. This is probably because many smaller animals do not survive the floods triggered by the storms. Salinated freshwater sources because of the storm surges also take their toll on the tigers. Every year, dozens of people fall victim to tigers. For this reason, and also because killing a tiger is regarded as a demonstration of courage, the animals are repeatedly shot by villagers [126]: a fatal cycle, which further disturbs the ecosystem's equilibrium.

The Bengal tiger is one of the most endangered species. Only about 100 to 150 animals roam the mangrove forests of Bangladesh. Several protection projects, often financed with foreign money, want to help save the population. In this context, a local resident asked me during my trip: "Why do you protect tigers and not people? Are tigers worth more than human lives in your country?" In southern Bangladesh, there are signs from aid organizations at many intersections pointing to species conservation and development projects. USAID, GIZ, Japan Debt Cancellation Programme, and Korea International Cooperation Agency—a host of national development aid programs and international NGOs—are present in the country. Yet, many people continue to live in abject poverty. The frustration of the man who approached me is understandable. The conflicts between wildlife protection and existential threats to village communities who live on forest resources are self-evident and difficult to resolve. The growing population is settling deeper and deeper into the Sundarbans, destroying the tigers' habitat.

A new large-scale project, the Rampal power plant in the southwest of the country, may prove fatal for the endangered species. The coal power plant is envisioned to meet the country's growing energy needs. It is financed with investments from India and located on the edge of a protected forest, threatening the already fragile ecosystem. The coal is transported to the power plant by cargo ships that make their way through the mangrove forests. Environmentalists believe that the transport and wastewater will also affect the protected area to the south, the Sundarbans World Heritage Site. The deterioration in air quality is also likely to have a severe impact on nearby villages. In 2016, the International Union for Conservation of Nature (IUCN) and UNESCO called for the construction of the power plant to be halted—without success, the power plant started operating at the end of 2022.

Local environmental groups demonstrated against the planned power plant, and activists were repeatedly victims of violence. People from the region where the coal-fired power plant is being built have already been displaced to make way for the construction work. This kind of displacement is also part of the climate crisis: people who are forced to leave their homes because of coal-fired power plants or mines.

Children and Climate Migration
Many children are born into the filth of the slums. When they are a little older, they sell cigarettes on the roadside or carry heavy water cans. Do they go to school? Not always. Some are malnourished. How often can these children be carefree? What does their future look like? In addition to poverty, these children "inherit" the burden of the global climate crisis, which will further worsen their life prospects if the powerful industrialized countries, the big polluters, remain inactive in the fight against climate change. Of the 30.1 million new internally displaced people worldwide due to weather events, nearly one-third were children in 2020 [127]. The UN Children's Fund (UNICEF) estimates that between 2016 and 2021, around 43 million children were internally displaced due to weather-related extreme events. These disruptive events often limit access to health services. Climate change can lead to food insecurity, waterborne diseases, such as diarrhea, and other health impacts, which for children can have lifelong consequences. For example, malnourishment for children can be correlated with lesser economic productivity in adulthood.

Globally, about 400 million children live in areas with tropical cyclone risks [128]. In a poverty-stricken settlement in the regional center of Khulna, we talk to a 14-year-old girl. When she was 7 years old, her father and grandparents were killed in Cyclone Sidr. Her mother could no longer care for her and went to Dhaka as a domestic helper. Since then, she has been living with her uncle's family. The girl can at least attend school, but she no longer has contact with her home village. "Everyone has died," she says.

The destruction of livelihoods and also the forced displacement from familiar surroundings can result in long-lasting trauma. However, psychological help is rarely available. When natural disasters strike, children and young people often lose their entire social network, and their schooling is interrupted. In some regions, children's aid organizations set up temporary schools, also in refugee shelters. BRAC (Bangladesh Rural Advancement Committee), one of the world's largest nongovernmental organizations, teaches nearly 750,000 students in Bangladesh. Some families tell me that they can only send their children to school because of the NGO's work—a cornerstone in the lives of young people that offers some hope for a possibly better future. During our visit, however, some fathers also report behind closed doors that they are forced to send their sons to madrasas, Islamic schools. Still, other children have to work instead of learning because their parents' income is not enough.

In this context, children with disabilities, orphaned children, and those living in extremely remote areas are in need of special protection and services. The UNICEF, together with the technology company Microsoft, developed the so-called Learning Passport. This is an online learning platform that provides learning material on laptops and tablets to children who are unable to attend regular schools as a result of natural disasters, pandemics, or wars. Some of the displaced children suffer from stigmatization. To prevent socioeconomic inequalities from growing, it is crucial to maintain educational opportunities despite adverse circumstances.

At one of our last interview locations, our driver looks uncertain as we drive into the village on a dirt road. Supporters of the Islamist Jamaat-e-Islami or Hefazat parties supposedly live here. Both organizations are associated with the most serious religiously motivated crimes, such as the brutal murder of secular bloggers and publishers.

We discuss the situation and decide together that it is safe enough to conduct the interviews. "But we absolutely have to be out of here before dark," the driver reminds us. It is already late in the day. If there is one thing I have learned on my travels, it is to listen to the advice of local people. They can better assess situations and evaluate risks accordingly.

The reason for our visit to the village is based on the information we received from migrants in other interviews about dramatic coastal and river erosion in the area. Because of the uncertain situation and the possibly conservative attitude of some villagers, the two men in our group get out first. Naoshin and I wait in the locked car. After a few minutes, they return. We are welcome and can join them. We meet the village headman, who greets us in a friendly manner and immediately leads us to a quarry. We quickly realize the full extent of the damage. A large area of fertile soil on the bank of a tributary of the Brahmaputra had broken away a week earlier and was washed into the riverbed. A family of several people was swept away with it. Onlookers come and look at the spot with us. The accident happened at night. It is risky to build settlements and farm close to the river, but all the surrounding areas are densely populated and people have few options.

Neighbors of the family who perished in the river live on an adjacent piece of land. We ask whether they will continue to live there, whether it is not far too dangerous. Their answer sounds hopeless: "We'll stay here until the river takes us away, too. Where else are we going to go?" Apart from the piece of land, the family has nothing to hold on to. So, they cling to this small plot, on which they hope to make ends meet for the time being. If they were to leave, they would probably have to struggle to make a living in one of the slums and eke out an existence under potentially more adverse conditions. The fight for survival marks their life.

Afterward, we want to conduct another interview with the village leader. A number of people are grouped around us. We explain why we are there and what we are investigating as scientists and whether he is willing to participate in an interview.

For scientific semi-structured interviews, the procedure is always very similar. You introduce yourself, the research project, and the intention of the project. Then, it is explained what will happen to the information subsequently; how it will be published, anonymized, or documented with clear names; and how long it will be stored. Participation is, of course, voluntary. Everyone also has the right to cancel the interview or not to answer individual questions. At the beginning, the consent of the participants is obtained in writing; otherwise, the interview cannot be used for research purposes. For people who cannot read, there is also the possibility of verbal consent, provided witnesses are present and it is recorded. After all these formalities have been clarified, we go through our questionnaire, but keep the possibility open to address issues that come up spontaneously and to ask additional questions.

The village chief agrees to the interview and describes the plight of the people living here. As he speaks, more and more people gather around us. Finally, a cluster of almost 60 people has formed. Those who joined us later and are standing in the very back, of course, did not hear our explanation at the beginning. They make themselves heard, expressing their anger and disappointment that the national government is not helping them, and shout indignantly into our conversation. The atmosphere becomes charged. We can feel the anger and hopelessness of the people. The village chief tries to calm them down, but the heckling continues and grows louder. Presumably, some residents think that we are affiliated with the government and that they finally can voice their justified frustration and criticism. As the mood is threatening to tip, I look at our translator and we decide by eye contact to end the conversation. So that this does not happen too abruptly, I ask two more short questions. Then, I thank those gathered around us for coming and participating in the discussion. Rashed translates, and the more I explain our request, the more relaxed the faces of the bystanders become. The crowd lets us go, and we are relieved as we make our way back to our vehicle. We did not meet any Islamist fanatics in the village, but we did meet people who let us feel their understandable existential fears.

We spent our last night in the south of Bangladesh in a pleasant guesthouse of an NGO in the middle of green thickets. A moment of peace after the exertions of the past days. But there is still a surprise: I open the closet, and a palm-sized spider crawls towards me. Although I do not suffer from Arachnophobia, I do not want to share a room with this animal. I quickly run downstairs and talk to our host. Laughing, he comes with me to my room, picks up the spider with his bare hand, and carries it outside. In view of other problems, the eight-legged creature is probably more of a joke to him. Calmed down, but with enough imagination to fantasize about more critters in my immediate surroundings, I pull my thin sleeping bag over my head and fall asleep dreamlessly and exhausted.

The way back to Dhaka goes smoothly at first. For hours, we drive over country roads; everywhere, I see adults with their children, people transporting their belongings, rickety rickshaws with chicken cages, and trucks loaded with goods, black oxen, vast amounts of building material or sacks full of food. Again and again, we drive past painted house walls on which the German or also the Brazilian flag can be seen. "All Support of Brazil" is written underneath, with a small Bangladeshi flag. The 2014 World Cup is coming up, and there are clear favorites among the foreign teams.

At the city limits of Dhaka, our journey comes to a standstill, the traffic chaos increases, we are repeatedly stuck in traffic jams for hours within the city. At one intersection, all vehicles seem to be caught in a Gordian knot. Police officers run across the intersection and try to solve the situation. In doing so, they target the weakest in the hustle and bustle, the rickshaw drivers. With swinging batons, they smash off the bicycle lamps of all the rickshaws, apparently as a punishment for violating traffic rules. Obviously, all road users are collectively responsible for traffic jam, but no police officer beats down the car hoods. The faces of the rickshaw drivers are contorted in pain, as if they had been physically hit. For them, a broken lamp means a serious loss of income, as they are not allowed to drive without lights. Moreover, many of the rickshaws are only borrowed. A destroyed lamp means starvation for them and their families. Even if they are undernourished and malnourished, cycle rickshaw drivers hardly appear in hunger statistics because physical exertion is not sufficiently included in the minimum calorie requirement that defines "hunger."

Cities for Climate Migrants?

How can cities deal with the growing problems of the climate crisis? What solutions are available in regions that have few financial resources? What steps for more climate justice need to be taken at the international level? These and other questions were at the core of Professor Saleemul Huq's work. Huq was the director of the International Centre for Climate Change and Development (ICCCAD) in Bangladesh, an institute that conducts applied research on the climate crisis and how to overcome it. He unexpectedly passed away in October 2023, but his legacy continues. A COP veteran, he attended all 27 international climate conferences in his lifetime [129]. I met him at some of these conferences, where he tirelessly brought the situation of developing countries into focus, spoke to government delegations, and gave eye-opening interviews. His charisma was one of great friendliness and openness, but his criticism of industrialized countries was razor-sharp. For many years, he addressed the issue of rural-urban migration due to climate impacts. Huq's credible thesis is that in Bangladesh, the vast majority of migrants will eventually make their way to the capital city of Dhaka, which is already densely populated.

Given this prospect, Huq strongly urges strengthening the resilience of regional centers and developing them into "climate-resilient, migrant-friendly cities." The latter is not always easy, as even long-established residents struggle with inadequate housing, high living costs, unemployment, and climate impacts. Special training opportunities or assistance for newcomers are therefore sometimes viewed with suspicion by the urban population, which is also affected by poverty. Some city administrators also fear that the more assistance provided to arrivals, the greater the influx. This is not necessarily the case. After all, especially when people migrate to ensure their survival, considerations such as labor market opportunities tend to play a subordinate role. Interview partners repeatedly told me that they would hold out in their villages as long as possible and even take considerable risks before even considering moving to a city. To prevent crisis situations from arising in the first place and to avert conflicts, Huq advocates programs that ideally benefit both sides. In rural areas, for example, children and girls in particular should be given better access to schools and training centers so that they can quickly gain a foothold in the urban labor market later on, should the need arise.

GIZ's Urban Management of Internal Migration due to Climate Change (UMIMCC) project in 47 slums in Bangladesh focuses on education and training to give migrants and local residents a chance at regular employment and better earnings. In addition, information on how to apply for government assistance and financial support is provided to migrants and other poverty-stricken people alike. Self-employed small business owners receive in-kind assistance. During the COVID-19 pandemic, GIZ, together with other aid organizations, provided additional funds to compensate for lost earnings. This measure was carried out in an unbureaucratic manner via cell phone transfer and supported women in particular.

The German Advisory Council on Global Change (WBGU), for which I worked as a research analyst from 2014 to 2017, also recommends promoting a polycentric urbanization in its report "Humanity on the Move - The Transformative Power of Cities." Strengthening medium-sized cities is intended to increase a country's systemic resilience to better cope with climate impacts and the resulting movements. Moreover, a decentralized urbanization may serve to counter growing inequalities between rural and urban societies.

The urbanization trend may lead to ever-larger megacities, but such structures face significant challenges because they require large amounts of energy to be generated outside the city limits and transported from there to the urban centers. They are also often difficult to manage and govern, especially when population growth is steep and job opportunities and housing supply are not keeping pace. In many cases, satellites of informal settlements form around the core city, where the state does not provide services and has limited enforcement powers. Last but not least, uncontrolled growth of megacities is fostering the emergence of megaslums with more than a million people, such as Orangi Town in Karachi, Pakistan, or Dharavi in Mumbai, India. Governments must therefore develop and implement strategies to promote small- and medium-sized cities. This includes enhancing the culture and identity of a place as well as creating economic opportunities through research or new industries. If all this were ensured, it would be possible to divert migration routes away from large metropolitan areas, where it is already becoming increasingly difficult to provide social housing. Future advances in digitization and 3D printing technologies could even support this polycentric spread of cities.

There are historical examples of polycentric urban formation that became cradles of progress and development, for example, the cities of Jena and Weimar during the Age of Enlightenment or the Emilia-Romagna during the Renaissance [130]. If it is not possible to foster a more decentralized development, megacities with ever increasing land values, competing with each other globally for resources and talent, may form a growing contrast to the devalued rural hinterland in the future, according to the WBGU's analysis.

One-Third of National Territory under Water

If a multitude of different measures would be set up to address climate migration in vulnerable countries today, this could also benefit countries where severe climate impacts may not be felt until later [131]. Progress in the field of disaster prevention and climate adaptation is urgently needed. In a pessimistic scenario that assumes growing climate impacts coupled with weak and unevenly distributed development, the World Bank projects nearly 20 million additional internally displaced persons in Bangladesh alone by 2050, which means that Bangladesh would account for nearly half of all future climate-related internal migration in South Asia [132]. Migration due to tropical storms is not even fully accounted for in these calculations. However, the warning signals are already abundantly clear; in 2020, one-third of Bangladesh was flooded during the monsoon season.

Despite all this, the country's development progress is remarkable. For nearly two decades, the economy has grown by a robust 6% annually. As a result of the successful fight against poverty, Bangladesh is expected to rise from the status of one of the world's least developed countries to the level of a developing country in 2026. One can only hope that climate change will not break this positive trend.

Because Bangladesh is particularly affected by climate change, it actively engages in the international discussions on this issue, for example, in the Climate Vulnerable Forum. This platform of South-South cooperation helps governments of countries such as Bangladesh and the Philippines, but also from other regions of the world such as from the African continent or Central America, to look for solutions and develop joint political positions for the international climate negotiations. The 48 participating countries are united by the goal of limiting global warming to 1.5 °C above the pre-industrial level and making their own economies emissions-neutral in the long term. Due in part to a lack of financial support for transformation in the energy sector, countries such as Bangladesh are still a long way from achieving this goal; the share of renewable energies in the electricity mix is less than 10%.

Although the lower temperature limit of 1.5 °C set in the Paris Agreement is basically out of reach due to decades of inaction, it remains of enormous importance to stay as close to this range as possible. This is because in highly exposed countries, such as Bangladesh or the Philippines, the limits of adaptation are becoming increasingly apparent. If storm frequency and intensity increase remain unchecked, people in many coastal regions will no longer be able to live safely in these areas.

In the past decade, a series of superstorms has already burned itself into the global collective memory: In 2015, Hurricane Patricia devastated parts of Mexico. It was the strongest tropical cyclone in the Western Hemisphere. Cyclone "Winston" destroyed swaths of Fiji's coastline in 2016 and was one of the most devastating tropical storms in the Southern Hemisphere. In 2017, Hurricane Irma triggered large-scale displacements in the Caribbean, leaving Barbuda in ruins. In 2019, Cyclone Idai caused severe damage in East Africa as it moved from the coastlines of Madagascar and Mozambique far inland to Malawi and Zimbabwe, where it destroyed crops with large volumes of water. In 2023, Cyclone Daniel became the deadliest tropical storm since Cyclone Haiyan, when it made landfall on Libya and caused thousands of deaths in the city of Derna. As the cities' dams broke, people perished in the floods, with estimations ranging between several thousand to 24,000 deaths.

Thousands, if not hundreds of thousands, of people are already displaced every year due to hurricanes, seeking refuge in emergency shelters and informal settlements—only to have to move again when the next extreme weather event occurs. Adaptation to climate change thus requires many to be mobile in order to ensure survival.

Thesis

Successive superstorms could depopulate exposed areas in the long term. Progress in disaster management in many Asian countries is a hopeful sign of higher resilience, but the advancements are competing with increasingly violent climate impacts.

Chapter 6
Fire in the Rainforest: Biodiversity Crisis in the Amazon Basin

Brazil: Guardians of the Amazon Basin ♦ Tireless Activists ♦ New Times and Old Challenges ♦ Lula III: New Times, Old Challenges ♦ Climate Change and Species Loss ♦ The Role of Diversity ♦ Planetary Health ♦ Peru: Three Vegetation Zones and Even More Challenges ♦ Lima: City Without Water ♦ The Relentless Boy: El Niño ♦ Weather Services at Crossroads ♦ Wandering Trees and Fortune Seekers ♦ The Future of Our Climate Niche

Looking at a satellite image of Latin America, the vast Amazon rainforest, which stretches across Brazil, Peru, Colombia, and, to a lesser extent, into neighboring countries, seems invulnerable. But as you zoom in, the open wounds of this majestic ecosystem immediately emerge: dead-straight roads stretch deeper and deeper into the forest. They pave the way for the timber industry, cutting into the dense greenery.

At present, the Amazon rainforest still covers almost half of Brazil's national territory—a huge, invaluable resource that is also home to enormous biodiversity. Globally, 80% of all species cavort in 20% of the terrestrial regions, predominantly in the tropics. The world's greatest diversity is found in the Amazon basin and the adjacent Andes. Unquestionably, this irreplaceable natural heritage must be safeguarded. However, in 2022, a study documented: The Amazon forest is dramatically losing its resilience, triggered by clearings and fires and aggravated by climatic changes [133]. This development must cause us extreme concern; after all, the Amazon rainforest is an important regulator in the entire climate system. It gives rise to "flying rivers" that transport water in the atmosphere into the South American precipitation system, where they currently still reliably ensure rainfall. However, due to deforestation, the flying rivers are at risk of drying out. The preservation of the forest is crucial for regional water cycles.

The tropical forest also stores large quantities of carbon that is absorbed from the atmosphere. However, after intensive observations and analyses, scientists are increasingly sounding the alarm, because in the past 10 years, the Amazon rainforest has emitted one-fifth more CO_2 than it has absorbed [134]. This worrisome turnaround is due to several interrelated dynamics. Foremost is a qualitative change in the forest, whose condition is increasingly deteriorating. Rising temperatures, fire, and forest fragmentation from road building and economic development are undermining the complex ecosystem and reducing its biodiversity. Biomass is decreasing even before the forest disappears completely.

The other big driver is deforestation, and with it the large-scale destruction of rainforest, which is increasingly making way for cattle ranching and serving as arable land for soy and other crops. Between August 2020 and July 2021, 10,476 km^2 of the rainforest was destroyed, the highest level in a decade [135]. Other areas of Brazil are also affected by the economy-driven destruction. In 2020, for example, the Pantanal biodiversity hotspot, a wetland southwest of the Amazon, saw the largest areas on fire since records began. Increasing drought there is causing small-scale slash-and-burn operations by ranchers to spiral out of control more quickly.

The Amazon rainforest is one of the tipping elements in the Earth system. A deforestation of 20–40% or a global warming of 4 °C could turn it into a savannah landscape. The forest is visibly approaching this point; by 2019, more than 17% had already been cleared, and since then, the destruction has continued, albeit some slowdown since the government of Lula da Silva took office in 2023 [136]. What is worrisome is that once such a tipping process has begun and becomes increasingly obvious, it can already be too late for rescue measures. Hence, there is an acute need for preventative action to avert this catastrophe because the complete destruction of such an important carbon sink and the disruption of the precipitation system on the South American continent would have global consequences. Instead of protecting the area, the opposite has been happening. Many regard the rainforest as a resource to be exploited for short-term profit.

Brazil: Guardian of the Amazon Basin

August 10, 2019, will go down in Brazil's history as a day of destruction. On this day, landowners in the Amazon rainforest decided to set out on a path of no return and start fires in the lush forest. The burned areas were to serve the agricultural economy [137]. Allegedly, the Ministry of Environment in Brasilia at the time under the presidency of Jair Bolsonaro was aware of the collusion of large-scale landowners but did nothing to stop the criminal activity. Instead of fighting the fire and establishing better protection structures, Brazilian government politicians from the extreme right of the former government of Bolsonaro instead resorted to "whataboutism." "Whataboutism" is the constant reference to other, similar problems in order to divert attention from their own responsibility. Thus, while the Amazon rainforest was going up in flames, other environmental problems were brought up to dilute the discussion: "What about the forests in Germany, Russia, Australia or Portugal? What about plastic pollution in the ocean?" From a species conservation perspective, it is of course not the same whether a tropical virgin forest is burning or a conifer monoculture in Europe is on fire, but this was irrelevant to the Bolsonaristas. The Brazilian government even boasted that, unlike in forest fires in other countries, no one has died in the brazilian fires so far. Diseases and long-term damage caused by the use of pesticides and deforestation agents in the run-up to the deliberately set fires were gladly concealed.

Only after massive criticism from abroad, for example, from French President Emmanuel Macron, who held the G7 presidency at the time, did the government send military into the area. This was purely symbolic, as the soldiers were neither trained nor equipped to stop the firestorm. Due to severe drought, the fires spread rapidly anyway. Even from space, the blazing flames were visible.

Less than a year later, on April 22, 2020, an internal interministerial meeting in Brazil discussed environmental deregulation. The meeting was scheduled to discuss the "Pró-Brasil" economic stimulus program. But it quickly became clear: The coronavirus pandemic was to be used to defuse environmental laws and give free rein to economic interests in the rainforest region, while public attention was focused on the pandemic [138]. The contents of the conversation only became known because a judge released them due to an investigation against the family of Brazilian President Bolsonaro.

In the recordings, Brazilian Environment Minister Ricardo Salles complains about the intervention of the courts in opposing unconstitutional environmental deregulation and destruction: "(...) it is very difficult. In this regard, I think the environment is the most difficult to adopt any infrastructural change. It's about normative instructions and regulations. Because everything we do is a problem in the judiciary the next day [139]." Former Environment Minister Salles already used the discussion of meat consumption and deforestation at the 2019 international climate change negotiations in Madrid as a provocation. On social media, he posted a photo of a giant beef steak with the inane caption, "A vegetarian meal to compensate for emissions."

His counterpart in the Ministry of Education, Minister Abraham Weintraub, who was previously a banker, not only mocked environmentalists at the interministerial meeting in Brasilia, but also insulted the people displaced by the land grabs of large landowners: "I hate the term 'indigenous peoples,' I hate that term. I hate it. The 'gypsy people.' There is only one people in this country. [...] It is the Brazilian people—there is only one people. They can be black, white, Japanese or indigenous, but they have to be Brazilian. No more of this business of peoples and privileges [140]." With this, he argues for the complete assimilation of indigenous peoples. The motivation behind this are economic interests in land granted to indigenous peoples, which they have inhabited for centuries. Weintraub has been in the negative headlines several times because of his provocative statements. Among other things, he called Brazilian constitutional judges "bums who belonged in prison"; and was finally forced to resign. Before that, he still used his diplomatic passport to leave for the United States during the coronavirus-related travel restrictions. But his resignation as education minister did not hurt him. On the contrary, soon after, he was given the post of Executive Director at the World Bank in Washington DC in the United States.

The Bolsonaro government's anti-indigenous and environmentally destructive agenda was evidenced by both words and actions. In 2020, in a speech to the UN General Assembly, he claimed that indigenous people, alongside peasants, were the ones responsible for fires in the Amazon region and that these fires were occurring in already deforested areas [141]. The Brazilian president revealed his racist sentiments long before he took office, when he told a journalist in 1998: "The North American cavalry were the competent ones because they decimated their Indigenous people in the past and today, they don't have this problem in their country [142, 143]." It is estimated that several million indigenous people lived in the Amazon before the colonization of South America, and their numbers declined drastically afterward—through murder, displacement, and the spread of disease. The extermination of indigenous peoples continued for centuries in various ways; for example, employees of the Brazilian Protection Agency for Indigenous Peoples (Serviço de Proteção ao Índio, SPI) poisoned entire tribes with arsenic-laced sugar, as reported in 1968 [144]. Today, it is mainly factors such as economic marginalization, environmental destruction, land grabbing, and climate impacts that threaten these groups and drive them from their territories. But there are also repeated and sometimes fatal acts of violence against indigenous environmental activists.

As soon as he took office in 2019, Bolsonaro took a series of measures to curtail the rights of indigenous peoples and open up forest areas to the interests of corrupt elites. For example, the powers of FUNAI (Fundação Nacional do Índio), the agency responsible for indigenous affairs, were restricted by decree. The agency was removed from the Ministry of Justice and placed under the country's much less powerful Ministry of the Family. This ministry was then headed by an evangelical pastor who believes Jesus spoke to her in a guava tree. At the same time, the president decided to assign the agency's authority to define and demarcate indigenous peoples' territories from other lands to the Ministry of Agriculture, which protects the interests of Brazil's powerful agricultural lobby. The demarcation of lands provided for in the Brazilian Constitution means some protection, since large entrepreneurs cannot easily farm or mine in these areas. These areas are exclusively for the use of members of indigenous groups.

Terms such as "indigenous groups" or "peoples" encompass extremely heterogeneous cultures, even within Brazil, with their own languages and cultures. The Cambridge Dictionary defines indigenous as follows: "used to refer to, or relating to, the people who originally lived in a place, rather than people who moved there from somewhere else [145]." How states recognize indigenous peoples and what rights they grant them can vary widely. Although indigenous people are estimated to make up only about 5% of the world's population, they speak 4000 of the world's

7000 living languages. Thus, they also maintain a cultural and social diversity that must not be lost.

Many land areas managed by indigenous groups have higher biodiversity than other areas [146]. The reasons for this vary depending on the region. For example, aspects of intergenerational use of land and cultural appreciation of natural resources may play a role. However, lack of access to industrial, exploitative forms of economic development can also contribute to preventing the overuse of resources. If land is placed under such strict protection that no use whatsoever is permitted, this may lead to conflicts of interest with indigenous groups. On the one hand, conservation means restrictions, also for local communities. On the other hand, it opens up prospects for future generations [147]. Only about one-fifth of all indigenous territories are subject to formal land rights systems. Many inhabitants are therefore in constant danger of falling victim to land grabbing and losing their ancestral lands. If the lands are "developed," i.e., cleared, this in turn exacerbates climate change, so that only a few people benefit in the short term, but many have to bear the consequences of greed and displacement in the long term.

Tireless Activists

In Brazil, many indigenous groups are guardians of the rainforest. A 2020 study shows that areas subject to formal land use titles and attributed to indigenous groups are significantly less deforested than other areas [148]. However, under Bolsonaro, and even under his predecessor, ex-President Michel Temer, no further formalization of land titles for indigenous groups has taken place. Because of this, resistance is growing from indigenous activists, who are fiercely fighting back in large protests in the capital Brasilia and with calls from the younger generation on social media.

One of them is Alice Pataxó, who campaigns for more climate and environmental protection on platforms like Instagram and educates people about prejudices against indigenous people. Together with like-minded people, she demonstrates against large agribusiness, which is pushing ever deeper into forest areas, and was also part of the movement against the previous Bolsonaro government. A bill introduced in 2022, referred to as "marco temporal", proposed to allow economic activities such as resource extraction in the territories of indigenous people. It would make the recognition of indigenous territory more difficult. The bill has since been contested. President Lula da Silva attempted to veto the bill, but he was outvoted by a majority in the Brazilian parliament. The case will be brought to the Brazilian supreme constitutional court, which had deemed the law unconstitutional prior to the vote. The vote in favor of the law after the court's decision can be interpreted as right-wing forces attempting to politicize the decisions of the supreme constitutional court. The Amazon is already an El Dorado for all kinds of illegal ventures, against which the government does little. This applies to the clearing of valuable tropical timber as well as to the operation of illegal gold mines, even within reserves.

If the law were to pass, the destructive forces would receive further impetus. Activists who oppose logging and agribusiness live dangerously. Indigenous people are repeatedly murdered, such as forest protector Jorginho Guajajara from the Guajajara tribe in the Brazilian Amazon, who was killed by loggers in 2018.

President Bolsonaro was willing to use also geopolitics to strengthen the Brazilian agricultural lobby. The war in Ukraine and the resulting price increases for fertilizers were used as a pretext to justify the overexploitation of indigenous land. He claimed that potash fertilizer for agriculture needed to be mined in the Amazon, even though the raw material potassium is available in much greater quantities in other parts of Brazil. The motivation for this is the deregulation agenda, which benefits the economic elites and deprives indigenous groups of their livelihoods.

Lula III: New Times, Old Challenges

Since the beginning of 2023, Luiz Inácio Lula da Silva from the Brazilian Workers' Party took office for the third time as president of Brazil. Bolsonaro lost the presidential election in 2022 by a thin margin, and his political fate was sealed by the perception of a chaotic government that poorly handled the COVID-19 pandemic. Moreover, his right-wing discourse did not resonate well in impoverished areas of the country.

It is worth pointing out that Brazilian democracy is formed by a wide range of parties represented proportionally in the parliament and directly in the major functions, such as president. The president, being cumulatively head of government and head of state, has to form coalitions and allocate functions in the government and in the ministries to reflect the political forces present in the parliament in order to achieve what is called "governability."

Lula formed a cabinet of 37 ministries and secretaries with some old names and some new allies. For the environmental and climate change ministry, Marina Silva took the lead. The choice carries a lot of symbolism given her origin, history as an activist for the protection of the amazon forest, and her engagement in the social justice questions connected to the environmental protection of the forest. Although Lula formed a progressive government based on his campaign proposals, the newly formed cabinet will have to negotiate pieces of legislation and budget allocation with a majoritarian conservative parliament that strongly represents the agricultural sector and the evangelical "moral agenda." The exports connected to the agricultural sector were an important factor in the Brazilian economic recovery, and that consolidated the sector as a political force in the congress.

In order to achieve a reduction in the deforestation, Lula will have to find the fine balance between social demands and economic interests reflected, for example, in the establishment of indigenous reserves, which the agricultural sector tried so hard to restrict in the Bolsonaro years. Some victories can be seen in the advancement of indigenous rights and reduction of deforestation, but only time will tell what Lula will achieve in his third term.

> **Photographs of Change**
> A tireless witness to destruction and documentarian of diversity is the Brazilian photographer Sebastião Salgado. Together with his wife Lélia Wanick Salgado, he published a breathtaking photgraphy book about the Amazon, *Amazônia*, after capturing the fascinating beauty of the rainforest on several trips to very remote areas [149]. His images speak of the grace and dignity of the people who live in this habitat. In the preface of the book, Salgado expresses the hope that it will not become a testimony of "lost worlds," but that the culture and nature of the Amazon will have a future.
>
> When I met Salgado and his wife together with Hans Joachim Schellnhuber in Berlin in 2015, it quickly became clear how deep the photographer's knowledge of the state of the Earth goes and how strongly the couple feels connected to people and nature. Salgado's best-known photographs include his images of displacement and flight, such as the genocide in Rwanda, Hurricane Mitch in Honduras, and the fate of displaced landless peasants in Brazil. In 2015, Wim Wenders' film "The Salt of the Earth" was released, which traces Salgado's life and work. Deeply affected by the horrors and destructiveness of humanity, which he had documented photographically on many of his travels, Salgado dedicated himself to the "Genesis" project by capturing the other side, that is, the still almost untouched nature. But he also made a very practical contribution to environmental restoration. At the end of the 1990s, he founded the Instituto Terra on his parents' farm in the Brazilian state of Minas Gerais with the aim of revitalizing the destroyed areas of the Atlantic rainforests by replanting them. The tree nursery attached to the institute can produce around one million seedlings a year. Within two decades, more than 2100 hectares of land have been reforested in this way. In addition, the institute supports surrounding farms in reforesting part of their land and diversifying their income.
>
> Some of Salgado's Amazon photographs were on display in 2021 during the international climate change negotiations in Glasgow. Organized by Brazilian activists, the exhibition in the rented "Brazil Climate Action Hub" pavilion offered a stark contrast to the official pavilion of the Bolsonaro government. With his eye-opening images, Sebastião Salgado shows the damage already inflicted on nature and, at the same time, captures the infinite beauty of our Earth—a beauty that must be preserved in its diversity at all costs.

Climate Change and Species Loss

"Which is worse, biodiversity loss or climate change?" is often asked. However, the differentiation in the question is inherently misleading because intact ecosystems need a stable climate and animal and plant species are part of the carbon cycle. Instead of talking about the extinction of species, scientists tend to talk about the loss of biodiversity, because this term encompasses not only the diversity of species but also, on a small scale, the genetic diversity within groups of species and, on a large scale, the entire diversity of ecosystems. Genetic diversity makes populations of species more resilient to change, and species diversity in turn can make ecosystems more resilient, for example, to the dynamics of climate change.

In its sixth Assessment Report, the IPCC (Intergovernmental Panel on Climate Change), or more precisely, the panel's Working Group II, draws attention to the consequences of climate change for biodiversity [150]. Research into these connections branches out into many areas, since biodiversity, as already described in the case of Brazil, is under pressure due to many different human influences. Habitat loss due to deforestation, toxic pollution, fertilizer and pesticide runoff into rivers and oceans, plastic pollution, soil sealing—all these are examples of the dangerous levers that modern man has been turning for decades [151]. That is why it is hardly possible to look at the consequences of climate change on biodiversity and thus also on future human habitats in isolation. In various ways, we are all part of the great experiment of the Anthropocene, the era in which humans have a decisive influence on the nature of nature.

There is no doubt that climate impacts destroy habitats, for example, because they become too hot for their inhabitants. Sensitive moorlands are drying out, and frequent forest fires and heat waves can further reduce or completely destroy already depleted species populations. Flooding and coastal erosion also threaten animals and plants, such as the Bramble Cay mosaic-tailed rat, which was endemic to Australia and has been considered extinct since 2016. Tropical corals, which only thrive in a certain temperature corridor, are bleaching and dying from rising ocean temperatures and CO_2-driven acidification. However, around 500 million people worldwide benefit from intact coral reefs, whether through fishing or tourism or as a protective barrier against storm surges. This ecosystem that has developed and existed over hundreds of millions of years is now tottering within a few decades. If it disappears, the livelihoods of countless people will also be at risk.

The previously finely timed change of seasons in the temperate zones of the Earth is also losing its balance due to earlier onset of warm periods and mild winters. Certain species can adapt to a limited extent, for example, some birds which now breed earlier. But these changes are probably not enough to guarantee survival in the face of increasing climate pressure [152]. In general, as global mean temperature rises, the risks for serious consequences for biodiversity are multiplying. From a warming above 1.5 °C, the risks of extinction increase significantly, especially for endemic species, i.e., those that only occur in certain ecosystems [153]. Thus, biodiversity also shows that every tenth of a degree of avoided warming counts.

The Role of Diversity

The crisis of the Earth system cannot be addressed by tackling climate change alone. A holistic approach requires taking the state of ecosystems into account. So far, however, the foreign, economic, and industrial policy processes that must converge to address the crises are rarely thought of comprehensively. The journalists Fritz Habekuß and Dirk Steffens conclude in their book on species loss: "It seems that most people today are less afraid of the end of the world than of the end of capitalism. How else can it be explained that the World Biodiversity Council's frightening extinction figures are met only with a shrug, while a looming economic recession causes panic? [154]." A question that follows is: How safe can we feel when numerous species are going extinct around us?

The relationship between climate, biodiversity, and cultural diversity is complex. What is abundantly clear, however, from the example of the Amazon, is that numerous "antidiversity" activities—such as the expansion of agribusiness and raw materials extraction in previously intact nature—are leading to a reduction in biological-ecological and social diversity. The loss of biological diversity is dramatic because with it we lose genetic information that has evolved over millions of years, from species that are in some cases still completely unknown to us. This is tantamount to indiscriminately erasing information from nature's hard drive. Moreover, the displacement of indigenous peoples to make way for plantations, gold mining, and cattle begins with their uprooting and often ends in their destruction. Forced migration and expulsion take on a completely new dimension when considering the destruction of cultural areas that have grown over centuries. Many groups have been wiped out before their culture and knowledge could even be recorded. On the other hand, indigenous people who pass on their knowledge of the environment to strangers frequently become victims of biopiracy. Unscrupulous companies patent the knowledge with their own name and profit from it without acknowledging or enumerating the originators.

The diversity of genes within species groups, which has evolved over millions of years, allows for natural selection to respond to change. You can put it this way: Loss of diversity weakens the resilience of ecosystems and that of human society to climate impacts, as well as those structures that could help resolve ecological crises. Unchecked environmental destruction ultimately means the destruction of diversity that we have not even begun to grasp.

Humans are now intervening with the natural systems on a global scale by way of massive pressures on biodiversity. Through the commodification of a plant or animal species for the international market, extensive homogenization is taking place. For example, although Peru has more than a thousand varieties of corn and potatoes, only a few are used for world trade. Thus, these varieties compete for acreage, while other species and traditional cultivation methods are marginalized and displaced.

This fatal dynamic is also unfolding in forests. Instead of understanding the forest as an ecosystem and preserving its resources in terms of medicine and research for future generations, it is being cut down and replaced by monocultures. This can be observed both in Brazil with regard to soy cultivation and in Germany with regard to the timber industry. In Germany, around 30% of all species are threatened with extinction. Since the problem affects practically all countries, international negotiations on the protection of biodiversity are being held to determine which general goals can serve the preservation of species. These include, for example, a possible long-term commitment to place half of all land under some form of protection by 2050 [155].

However, just like international climate protection, global conservation efforts also show only few measurable successes. Despite various agreements, effective implementation of measures has failed so far, also due to lack of funding. How local species protection can become relevant globally is reflected not least by the wildlife trade in China, which was most likely the trigger of the COVID-19 pandemic. Despite this recognition, too little is being done to stop habitat and wildlife population loss or its illegal trade. However, healthy ecosystems are essential for human survival [156].

Planetary Health

How closely human health is linked to the health of the Earth system is captured by a still relatively new branch of research called "Planetary Health" or "One Health." The basic understanding of the concept is that human health is dependent on a healthy environment, such as clean air and water, fertile, unpolluted soils, or climatic stability. It also often involves the complex interactions between climate impacts, the resulting damage to ecosystems, and the consequences for human health. Some health risks begin already with the burning of fossil fuels, which are directly responsible for local health risks due to air pollution. WHO estimates that 4.2 million people die each year due to outdoor air pollution and 3.8 million due to indoor air pollution [157]. Some scientists even estimate that more than eight million people lose their lives prematurely each year due to outdoor air pollution from fossil fuel combustion alone [158]. While almost every inhabitant of the globe (99%!) is affected by poor air quality, some regions and especially certain cities such as Cairo or New Delhi have dramatic levels of pollution [159]. Moreover, the indirect effects of fossil fuels continue to have an impact for a long time. CO_2 remains in the atmosphere for several hundred years.

Heat deaths, the greater spread of communicable diseases such as malaria, injuries from storms, and even mental illness are among the many health consequences of climate change. According to the prestigious Lancet Countdown, the climate impacts on health are growing [160]. Directly or indirectly, they also influence migration decisions [161]. Slowly emerging risks from heat or air pollution are often accepted over longer periods of time, meaning that people tend to remain in face of gradually increasing risks. For example, life-threatening storms, however, trigger sudden movements. Some groups, such as people with disabilities or children, rely on special protection during extreme weather events. These considerations need be included in evacuation plans and disaster management.

Migration can improve or worsen access to health services. Sometimes, more health-care facilities are available, but the costs are prohibitive. During migration, people experience health risks to which they were previously not exposed. Big cities in Brazil are places of arrival for many migrants from rural areas and other South American countries. But those without a good education face a hard life. Displaced people from a wide variety of backgrounds live in informal settlements on the outskirts of the Amazon city of Manaus: Haitians, indigenous people, and Venezuelans. Regularly, authorities chase them away from their poverty-stricken settlements.

Many of them arrived in Brazil via a wide network of bus routes across the Latin American continent. For example, Brazil and Peru are connected by the world's longest bus route, which takes about 100 h to travel from Rio de Janeiro to Lima, the capital of Peru. The route was often used by Peruvian migrants who came to Brazil from rural areas in the early 2010s to seek work in the big cities.

Peru: Three Vegetation Zones and Many Challenges

Dusty, dry coastal highlands, majestic Andean mountains with glacial lakes, and tropical rainforest—the diversity of Peru's landscape makes a lasting impression on visitors. The challenges of climate change in very different vegetation zones can be seen in the country. In the research project East Africa Peru India Climate Capacities (EPICC), funded by the German Federal Ministry for the Environment, we set out together with Peruvian scientists to develop research questions on climate impacts on land and water management and to analyze the effects of climate change on internal migration. My colleague Dr. Jonas Bergmann researched the multifold connections between climate impacts and migration in Peru and what consequences they bear for the well-being of affected individuals [162].

In the Peruvian rainforest, Bergmann conducted interviews with people whose livelihoods were under massive threat due to riverbed encroachment. Two small villages in the province of San Martin in the north of the country were particularly affected. In one case, after many years, the approximately 150 inhabitants managed to acquire land and thus create a small livelihood, a minimum of security. Only to a very limited extent could they rely on outside help from the state or nongovernmental organizations: "Everything we have, the little we have, we owe to ourselves," one resident told him [163].

The other village community, with about 700 indigenous inhabitants, is still holding out in the flood-prone area. Because they do not have enough money to buy land, they applied for government aid to relocate in 2015. Although they were promised assistance, nothing happened, even after local authorities officially declared the area uninhabitable. The structurally disadvantaged people whose shacks stand in close proximity to the river live in constant fear of the next flood [164]. They have no real possibility to enforce their rights.

The Peruvian government has made resettlement from danger zones an important instrument of disaster reduction. Nevertheless, there is a lack of funds, capacities, and often political will to help strongly affected settlements. Marginalized groups such as indigenous people receive little support and are even more exposed than others to deteriorating environmental conditions.

But there are nongovernmental organizations that try to alleviate the suffering of poor rural populations by opening up new livelihood options. The Lima office of Caritas, for example, showcases projects it supports throughout the country. The Catholic aid organization works for the empowerment of communities. For instance, they enable small-scale coffee growers to sell their products on a larger market by providing technical and commercial trainings. Caritas also helped a group of pineapple farmers. By bringing together several farmers and investing in better equipment and organizational training, they were able to almost double their harvest. Instead of pressing the fruit for juice, it is now sold as a whole. This increases sales and ultimately creates new jobs. The Peruvian arm of Caritas also runs projects to minimize climate risks in addition to the classic rural development measures. For example, early warning systems for extreme hydrological events have been established, and information on community risk management has been disseminated in workshops.

In the following years, in collaboration with the International Organization for Migration (IOM), we develop a risk report that summarizes existing knowledge about climate impacts in Peru and their effects on internal migration in the country [165]. The climate risk projections are worrisome, to say the least. For example, our analysis revealed that the country could be affected by three distinct and potentially parallel unprecedented hazards if global warming approaches 4 °C. Each geographic zone is affected by different risks.

First is the transgression of thermoregulatory limits in the Amazon. Such limits can be identified, for example, by the so-called wet bulb globe temperature. This unit of measurement was developed by the US military to prevent heat deaths during training missions. It includes variables such as temperature, wind strength, solar radiation, and humidity. When global warming leads to extremely hot and humid conditions, people cannot stay outdoors for extended periods of time because they can no longer regulate their body temperature by sweating. If such extreme weather conditions occur on more than 200 or even 300 days a year, these areas are very likely to become uninhabitable. This would particularly affect the tropics, such as the Amazon region, should global emissions continue to rise [166].

Second is the virtually complete melting of the glaciers in the Andes Mountains, which would limit the supply of freshwater [167]. Already today, the condition of the glaciers is bad, as they are increasingly losing mass. Since the 1960s, there has already been a glacial retreat of around 40%, and just recently, the ice giants have been losing more and more surface area [168]. Initially, melting ice means an abundance of water. Glacial lake outbursts can lead to life-threatening floods, as in the case of the Andes village of Huaraz (see Chap. 2). However, once the peak of melting has passed, drought follows. Until now, natural seasonal glacier melt provided rivers with additional freshwater, especially in the summer months. If glaciers cease to exist, this water will be missing in the summer. Such a disruption in the system of tropical glaciers can occur in just a few decades, while some individual glaciers have already disappeared.

In a high-emission scenario, all glaciers would disappear by the end of the century. If temperatures stabilize at 1.5–2 °C, perhaps 20% of the sensitive ice masses could still be saved. More comprehensive protection of the Andean glaciers is no longer possible; too many greenhouse gases have already been emitted. This is also due to feedback effects, i.e., the lower reflectivity of dark, ice-free areas as compared to light, ice-covered areas. This reduced albedo effect drives warming, leading to further ice melt. Nearly one-third of Peru's population lives in the mountainous area and is likely to be directly or indirectly affected by the consequences of glacier melt.

Third is the devastating effects of extreme El Niño events on Peru's coasts, which will lead to flooding and a dramatic reduction in fish stocks. This will be discussed in more detail later.

These three scenarios not only pose significant threats, they also carry far-reaching implications for migration and displacement. Even if a fatal 4 °C scenario can be avoided: Already at 2 °C, significant damage would occur, forcing people to relocate. It is true that Peruvians have been migrating for hundreds of years to avoid natural hazards. But today, the country faces dramatic change that could both reinforce and alter existing migration patterns [169]. Rural-urban migration is increasingly influenced by climate change, and regional centers such as the Andean city of Cusco or Arequipa in the south of the country are common destinations. But even more people are seeking shelter in the megacity of Lima, which faces significant infrastructural challenges.

Lima: City Without Water

North of the Atacama Desert lies Lima one of the driest major cities in the world, with almost ten million inhabitants. Since the middle of the twentieth century, when the capital of the Andean country only had about one million inhabitants, the population has grown rapidly. Migration from the bitterly poor rural areas was and is one of the main drivers of this ongoing development. People still settle around the dusty capital in extremely barren landscapes. Open sewers, unpaved gravel roads, and lots of street dogs mark the surrounding. On the steep slopes, countless shacks crowd closely together. Many of the residents come from areas where extreme weather conditions have led to ever-increasing poverty. Two million people live here without access to running water. They depend on the services of expensive water delivery trucks.

Lima was not always so arid. In pre-colonial times, before the violent appropriation by the Spanish conquistador Francisco Pizarro, it had abundant water and forest resources. It was not until the bloody colonization of Peru that a centuries-long exploitation of resources began, eventually contributing to the deforestation and desiccation of Lima, which was also aided by climatic changes. A probably previously species-rich, vibrant cultural landscape eventually developed into an urban wasteland.

A group of scientists around the Lima-based International Potato Center wants to stop and reverse this trend [170]. As part of a global ideas competition organized by the US Rockefeller Foundation, they are proposing three interventions to transform Lima into an eco-metropolis and give the poorest parts of the city's population access to fresh food. Their plan is called "Lima 2035" and includes a series of scalable innovations.

In doing so, they are building on the success of smaller NGOs. The Movimiento Peruanos Sin Agua (Peruvians without Water Movement) has already begun to use simple means to remove water from the air [171]. It takes advantage of the fog that often lingers seasonally around Lima by setting up "fog catchers," fine plastic nets attached to simple wooden sticks that collect moisture trapped in the air in the form of condensation. This results in up to 400 L of water daily in some places. This water can be used to grow plants and trees, which in turn can regenerate the water balance in the long term. Further filtration, for example, by the reverse osmosis process, turns it into drinking water that can be bottled and taken home. The advantages of fog collectors are undeniable. Because they are very inexpensive and do not require electricity, they can serve those who barely have enough money to make a living. If applied consistently, it is possible to imagine that the method of collecting water from fog could revitalize entire regions. The principle of fog capture is also used by Chilean architects Susana Ortega and Alberto Fernández, who are creating designs for tall spiral towers with large woven surfaces.

The "Lima 2035" plan also includes urban gardening. Not only private houses are to be used for this purpose, but also the threatened archaeological sites from the early Inca period, so-called huacas (the word huaca comes from the ancient Inca language Quechua and means place of ceremony). These huacas are often just lovelessly fenced in due to a lack of money and are basically left to decay. The plan is to carefully integrate gardens into them, creating both food sources and spaces for encounters. Migrants who settle in Lima would then have the opportunity to plant small green oases that can support food security. As they often come from rural areas of Peru, they bring with them significant agricultural knowledge. Out of more than 1300 applications, the proposal "Lima 2035" was chosen as 1 of the 10 winners of the Food System Vision Prize, established by the Rockefeller Foundation. A huaca is now to be converted for vegetable cultivation on a trial basis, and eventually new life may grow out of the neglected archaeological site.

The Ruthless Boy: El Niño

Today, in Lima's more affluent neighborhoods, a closer look reveals infrastructure that protects against flooding. Because in years when an El Niño occurs, the dry metropolis is hit by extreme rains. An El Niño event is a change in ocean circulation that occurs at irregular intervals and results in a warming of the surface temperature of the eastern Pacific Ocean, precisely where Peru is located. El Niño ("The Boy") was named by Peruvian fishermen after baby Jesus, born in December, because the phenomenon usually occurs in that month. On Peru's coasts, it causes a drop in fish yields because nutrient-rich cold water does not reach the surface, disrupting food networks in the sea. It also brings flooding due to extreme precipitation triggered by the higher temperatures.

But that is not all. Depending on the strength of the El Niño event, extreme weather conditions develop in very different regions of the world. While floods occur in one part of the world, droughts can occur elsewhere, for example, in northeastern Brazil, the Amazon region, South Africa, and northern China. On the Indian subcontinent, severe El Niño events can lead to a drop in agricultural yields if the monsoon rains, which are essential for survival, are delayed or even fail to materialize.

Predicting El Niño events is a science in itself and fills entire journals. It is enormously important because when it is known that an El Niño is building up, precautions can be taken, such as stockpiling food reserves and setting up distribution centers. That is why one group in our EPICC project was also looking at early forecasting of such events and the onset and retreat of the Indian summer monsoon [172].

The cold sister of El Niño is La Niña. During a La Niña event, the surface temperature of the eastern Pacific Ocean drops. This brings comparatively favorable weather conditions for Europe, but it leads to more tropical cyclones in Asia, more droughts in western Latin America, and flooding in eastern Latin America and southern Africa. In many ways, and very simplistically described, the effects of La Niña are the reverse of an El Niño, in the wake of which they often occur. The El Niño and La Niña events affect climate so strongly that with the latter, global annual mean temperatures drop, while with El Niño, they skyrocket. Climate change could cause more extreme El Niño events to occur [173]. However, in some places, it also counteracts the effects of the circulation anomaly. For example, warming can inhibit rainfall events that normally occur due to El Niño, as has been observed in Chile [174].

Historian Mike Davis, in his book *Late Victorian Holocausts: El Niño Famines and the Making of the Third World* [175], describes the serious consequences of severe El Niño events at the end of the nineteenth century. Between 1876 and 1879, 1889 and 1891, and 1896 and 1902, extreme droughts and crop failures occurred across much of the globe. According to current research, this period was marked by extreme El Niño events. Climatic extremes were followed by outbreaks of malaria, cholera, bubonic plague, and other communicable diseases that struck an emaciated population. In combination with colonial exploitation, which forced agricultural workers to grow export crops rather than feed themselves, between 30 and 60 million people died in China, India, and Brazil alone. Mind you, there were less than two billion people on Earth at the time, which makes the relative magnitude of this destruction of human life even more frightening.

Davis describes the violence of starvation in his book as follows: "imperial policies towards starving "subjects" were often the exact moral equivalents of bombs dropped from 18,000 feet." Thus, the historian draws a comparison in his text between the famine-related deaths and the victims of nuclear bombs in World War II. He later underscores why famine is not merely the result of a natural disaster but, more importantly, the result of conscious decisions about the distribution of food and wealth. In his book, Davis depicts the grotesque economic inequality in the British Empire. He describes colonial masters surrounded by servants, reveling in wealth, while the masses died miserably of starvation, including numerous children. In the prologue, he refers to US ex-President Ulysses Grant, who in 1877 took a vacation trip through drought-stricken Egypt, India, Burma (present-day Myanmar), China, and Japan, with lavish feasts in each country while the population had almost nothing to eat.

The wealth of the elites perhaps only appears so inconceivably obscene in retrospect. Today, inequality spiraling out of control is simply accepted. The wealth of the rich in the globalized world economy has taken on unimagined proportions. The development aid organization Oxfam published shocking figures in 2017: The eight richest individuals (all male) owned as much as the poorer half of humanity—3.6 billion people at the time of publication [176]. The coronavirus pandemic did not break the trend of capital accumulation—on the contrary. The ten richest men in the world doubled their assets from $700 billion to $1.5 trillion between 2020 and 2022 [177]. With his historical analyses, Davis reveals that human suffering caused by natural disasters can be further potentiated by exploitative market mechanisms. As such, this meticulously researched work also points at how to deal with climate change.

Extreme inequality now operates on at least two levels: First, directly through an economic system that makes some people incredibly rich, so rich can even catapult themselves into space, while hundreds of millions of people exploited by this system cannot even fulfil their most basic needs. Secondly, the excessive consumption of the richest strata of the population leads indirectly to the degradation of livelihoods for all people but hits the poorest populations hardest and fastest. As the negative externalities of this consumption are not reflected in pricing, these damages are not even being paid for.

Weather Services at Crossroads

Lima's old town is still characterized by its colonial heritage, the appropriation by the Spanish conquerors. The city's colorful markets offer everything from local art to Chinese-made knickknacks. Nearby hip restaurants open their doors to tourists and the local elite, who are undeterred by the prices. In Lima's business district, on the other hand, rows and rows of chain restaurants and fast-food joints have taken over.

Squeezed in between some high-rise buildings is the old structure of a government organization that has been leading Peru's weather forecasting and climate modeling for decades, the National Service for Meteorology and Hydrology, SENAMHI. At the time of our collaboration on the EPICC project, Ken Takahashi, a world-renowned Peruvian El Niño researcher, was president of the Peruvian Meteorological Service. With Takahashi and his collaborators, my team and I had passionate discussions about the possibility of predicting El Niño and more generally about the impact of climate change on the people of Peru. Notably, at SENAMHI, many young women work in hydrology or meteorology, which is unfortunately still rather unusual in many other places.

The weather service has been dealing with another phenomenon for some time, the "coastal El Niño" that occurred in Peru in 2017. This one is not a true El Niño, because it does not result from a change in the circulation of the eastern equatorial Pacific [178]. But its effects can be just as devastating. The 2017 coastal El Niño is believed to have occurred in part due to a combination of very warm water temperatures along the coast of Peru and high humidity over the Andes [179]. The extent to which climate change is related to this phenomenon is still a matter of scientific debate. The example of the coastal El Niño shows that there are still many unanswered questions about the exact impacts of climate change on Peru and its people. In addition to the intense academic discussions, which are conducted in the spirit of transdisciplinarity, the hope of being able to lay the foundations for better climate adaptation through scientific analyses binds us together in the project.

Gene Banks: Reinsurers Against Biodiversity Loss?
The "International Potato Center" is located in eastern Lima. What sounds like a mouth-watering restaurant is part of a group of global research institutions that preserve crop seeds and genetic material for future generations under the auspices of the CGIAR (Consultative Group on International Agricultural Research). The common mission is to promote human development and fight hunger by reaping the benefits of biodiversity. Despite the immense abundance of food crops, only a vanishingly small number of varieties are used for much of the global food production. As a result, varieties that produce high yields are primarily grown, even if this is only possible with mass use of pesticides and fertilizers, which pollute the soil. This concentration on a few varieties displaces other "old" varieties that might cope better with certain pests or even extreme weather conditions. Seed banks have been created to provide a backstop against the complete extinction of important food crops. But not all plants can be stored as seeds. Therefore, these varieties are stored either by continuous cultivation in the field or as small plantlets in glass tubes (in vitro) and cooled so that growth is slowed and longer-term storage can be realized. This is also done at the International Potato Center, which conducts research on food security issues. The stored seedlings and

seeds are meticulously inspected for pests, fungi, and diseases. The center regularly passes on seeds and seedlings to farmers, because the best insurance is cultivation by the population.

There are also seed banks in Europe, for example, the Norwegian-funded Svalbard Global Seed Vault in Spitsbergen. The vault for crop seeds from all over the world is built into the permafrost. The intention is to be able to breed certain plants again in the event of extermination by disasters, such as major fires or chemical accidents. The vault is also an important reinsurance against the ongoing destruction of biodiversity. Over one million seed samples are stored in the state-of-the-art facility, which is cooled down constantly to $-18°$. Additional generators ensure cooling even during power outages.

But there was one event the seed bank in Spitsbergen was not prepared for: global warming. High outdoor temperatures and melting permafrost caused water to seep into the entrance gate of the facility, which was built into a mountain, for the first time in 2017. Alarmed scientists pointed to future problems caused by warmer weather conditions. Extensive upgrades worth tens of millions of dollars were approved to continue protecting the natural heritage at the seed bank.

Wandering Trees and Fortune Seekers

Peru has created a total of 15 national parks and a number of reserves since the 1960s. This benefits the stability of ecosystems. National parks are subject to the strict protection regulations of the International Union for Conservation of Nature (IUCN), which also sets guidelines for the management of such parks. The largest facility of this kind in Peru is Manu National Park in the southeast, covering an area of 1.7 million hectares. Conservation efforts are at odds with commercial interests in the park and its foothills, the so-called buffer zones.

Pressure on one of the most biodiverse spots on Earth is coming from several directions. New roads provide access to previously untouched areas and are gateways for logging and resource exploitation. Gold is mined largely illegally in Madre de Dios, the region where the national park is located. Mercury is often used to separate gold from the sandy water because it amalgamates with the precious metal. In consequence, mercury enters directly into rivers and poisons not only the impoverished fortune seekers, but also other people who live along the river and of course the surrounding wildlife. The illegal gold rush picked up steam during the 2008 financial crisis, after gold prices rose on the world market and when many people lost their jobs. Peru is one of the ten largest gold exporters in the world. Indigenous people who catch fish from the contaminated rivers fall ill as a result of mercury poisoning [180, 181]. Breastfeeding mothers pass the toxic substances on to their babies, who can suffer brain damage as a result [182, 183].

In 2015 the Peruvian government had to declare a state of emergency because tens of thousands of people were struggling with symptoms of poisoning. It sent military forces into the area to take down illegal gold panning operations. Nevertheless, the damage was already well advanced; many large areas had already been contaminated and cleared, and organized crime had spread into previously uninhabited areas. Manu National Park was also affected via its waterways. Its vast area is difficult to control, especially when few resources are available.

In the gold panning facilities, which are usually operated by criminal gangs, migrants from rural areas also try to make a living by digging for the precious metal. Paradoxically, people who have become impoverished due to environmental changes and who have to look for new sources of income also become accomplices of a comprehensive destruction that robs future generations of their livelihoods. The big money, however, is made on another level, at the latest when thousands of kilometers away the dirty gold lands in form of freshly polished rings at international wedding fairs. Even more problematic is the storage of thousands of tons of gold as national reserves. The United States has the largest gold reserves with over 8000 tons, followed by Germany and Italy. Extracting this gold has caused significant environmental damage and mining continues. Fueled by the Russian war of aggression against Ukraine, the price of gold has again been sharply rising.

On the edges of the national park, oil and gas are also sought after, and in some cases extraction has begun. The multinational oil company Shell even carried out tests in the national park itself in order to prospectively develop new sources. This exploration stands not only in stark conflict with local nature conservation, but also with the Paris Climate Agreement. The majority of the already known reserves of oil and gas can no longer be burned, in order to keep within the carbon budget of the warming limits.

Last but not least, climate change itself is also taking its toll on this fascinating national park. This is described in all clarity by the American science journalist Elizabeth Kolbert in her chilling book *The Sixth Extinction*. Together with Professor Stefan Rahmstorf, I met her on the Telegrafenberg in Potsdam, where she told us about her current work and asked us about our research. In the conversation, her deep understanding of the interrelationships in the Earth system becomes clear, as does her passion for the subject. From Manu National Park, which she visited herself with a research group from Wake Forest University, she reports on how the sensitive plant species in the deep green adapt to climatic changes.

Like humans and animals, trees also "migrate" to adapt to the new temperatures, primarily uphill to higher and cooler altitudes. The green giants do this via their reproductive pathways, i.e., by spreading their seeds in the wind or by forming seed-bearing fruits eaten by birds or bats. The seedlings survive and grow only in the temperature corridor of their climatic niche. Thus, the species move successively up the mountain slopes [184], in Manu demonstrably 2.5–3.5 m per year [185].

But some trees are not migrating fast enough to protect themselves from extinction. That is because higher elevations are also getting warmer, and certain species that live in natural niches are losing the environment they need to thrive. While national parks are important pillars of species protection, "in contrast to, say, a logging crew climate change cannot be forced to respect a border. [...] And with so many species on the move, a reserve that's fixed in place is no stay against loss," Kolbert writes [186]. Therefore, it is important that conservation efforts extend beyond the national park terrain, for example, by establishing corridors between protected areas to allow animals and, indeed, trees to "migrate." All in all, the way we as humans live must be brought into balance with nature in a holistic way—a mammoth task [187].

The Future of Our Climate Niche

Compared to the trees, plants, and animals of Manu National Park, humans are more mobile and adaptive. Technological and structural innovations have even enabled us to endure climatic extremes—we were thus able to spread out over the entire globe. Nevertheless, human settlement as well as agricultural activities are concentrated in a specific temperature corridor. We, too, occupy a climatic niche.

In the seminal scientific article entitled "Future of the Human Climate Niche [188]," an international team of authors describes that the climatic niche for human development is between 11 and 15 °C annual average temperature. This temperature range is where humans have increasingly settled over the past 6000 years. This is where the greatest agricultural yields and economic developments are achieved. However, climate change is shifting these temperature zones within a few decades. Climatic conditions of over 29 °C annual mean temperature, found today on less than 1% of the Earth's surface, may affect nearly one-fifth of the area in the future if global warming is not limited, the authors say. So far, such extreme conditions are mainly found in the Sahara. One-third of the world's future population would be exposed to these temperatures if people do not migrate. But even if emissions are rapidly reduced and global warming is limited to below 2 °C, about 1.5 billion people would still live outside the 11–15 °C climatic range of *Homo sapiens*. Since other, cooler areas would be more suitable for human development, it is conceivable that migration would enable adaptation. But unlike migrating trees, migrating humans face national boundaries which they may not be able to cross.

Thesis

The interplay of species extinction and climate change is pushing humanity and its habitat toward an unprecedented crisis. Only migration will allow certain species to survive.

Chapter 7
Climate Crisis in Germany and Switzerland: From the Halligs to the Alps

Reconstruction in the Ahr Valley ♦ Unheeded Warnings ♦ Fossil Disaster Management ♦ From the Frying Pan into the Fire: Extreme Precipitation and Climate Change ♦ Displaced by Coal Mines ♦ Hot Times ♦ Threatened Alpine Idyll ♦ German Coastlines Under Pressure ♦ Near and Far Effects of Climate Change

"Coffee and cake" is a German tradition that my Brazilian husband no longer wants to do without. So, we sit in a café in the sun and enjoy the baked goods and the picturesque surroundings: flower boxes on the windowsills of old half-timbered houses, greenery everywhere, a small town out of a Germany travel guide. A brook babbles beside us, a few ducks fight over a crumb of bread. We are in Bad Münstereifel, barely 3 weeks before the catastrophe.

On July 14, 2021, flash floods triggered by extreme rainfall wreak havoc on an entire region. Overnight, the water levels rise from less than 1 m to more than 7 m. Exact figures are hard to get, as many measuring stations have been swept away. The Ahr and its tributaries turn into raging rivers, bringing death and destruction. Moreover, 135 people die in the Ahr Valley alone. In total, more than 200 people perish. The enchanting old town of Bad Münstereifel is completely destroyed. After the water masses have receded, the full extent of the catastrophe is visible. The streets are hardly recognizable. Mud and rubble dominate the cityscape; many people lost all their belongings in the floods; some mourn the loss of family members and friends.

About 10 months after the disaster, I visit Ahütte in the Rhineland-Palatinate municipality of Üxheim, a place that was not the focus of media attention but was heavily affected by the tragedy. A narrow road leads from the village center to the house of Renate Petry and Ulrich Schulz, who run a small guesthouse there. The couple lives in and with the nature of the volcanic Eifel. While Ulrich Schulz runs an occupational therapy practice out of town, his wife manages the house, guesthouse, and farm. Since the flood disaster, reconstruction has dominated their everyday lives. The couple's home is in a lovely setting, lined with mixed forests that wrap their dense greenery around the area. They used to get most of their food from their own cultivation; they raised chickens and also used their nature-oriented way of life as the core brand for their guesthouse, which is called "Behind the Island." Over many years, they painstakingly built up their garden and yard. Guests could find peace in this oasis, hike, eat freshly laid eggs, and simply enjoy the landscape. A small mill wheel supplied the farm with electricity. But the brook that drives the water wheel is a tributary of the Ahr and runs in the immediate vicinity of the house.

Floods are nothing completely unusual for the family. In the past, too, the stream has burst its banks, and water has made its way into the farm garden. But Renate Petry does not foresee anything good when she hears the extreme rain warning. On the morning of July 14, 2021, the day on which everything was to change, she drives once again to her parents-in-law, who live just under half an hour away from Üxheim. This time, the farewell is very detailed: "I told them that they should keep calm in case of flooding. There was a premonition that it could be bad this time." Back in Üxheim, she phones her husband, who is at work, and discusses the situation. The two decide together that Ulrich Schulz should make his way home as quickly as possible. If there were any road closures due to flooding, he would not be able to reach the house. The situation changes dramatically: "At noon there was still nothing to notice - at 2 p.m. it was clear: something is coming our way. And it's coming much faster than we're used to."

Ulrich Schulz arrives just in time. A short time later, landslides block the access roads. Since the village lies upstream of the Ahr, the water levels rise rapidly in the late afternoon. Around 5 p.m., water enters the cellars. "The water came with furor, and we realized we had to retreat," reports Mrs. Petry. Both had experience with lighter floods and thus intuitively knew that they should not start moving items to safety outside or in the basement now. Again and again, many people are killed as a result of such heat-of-the-moment decisions. Meanwhile, while the basement is filling up, water is also seeping through the walls into the first floor. "Our house was completely flooded, from all sides." Next, the power and telephone fail, and Ulrich Schulz's cell phone sinks into the floods. From this point on, the couple is completely cut off from the outside world and exposed to the forces of nature.

On the terrace, the water rises more than a meter. Everything they have built up is now at stake. Miraculously, the large double door that separates the kitchen-living room from the outside does not give way. For this reason alone, the damage to the interior of the house is limited. But in the morning, after a never-ending night of flooding, the extent of the destruction is revealed on the exterior of the courtyard. "I will never forget this sight," Renate Petry describes the situation. Virtually, everything is broken; the water mill and sewage treatment plant are destroyed; the self-supply garden and the chicken coop are engulfed by mud and water. Despite her despair, Mrs. Petry realizes the very day after the flood that they must accept this catastrophe in order to somehow continue living. But her thoughts are not only about the economic damage.

The hours and days after the flood are marked by agonizing uncertainty. They do not know whether Ulrich Schulz's parents survived. Both are over 80 years old and live in Insul, near the destroyed town of Schuld in the district of Ahrweiler. Communication networks have collapsed in large parts. Only gradually does information reach the people affected. It is not until the second day that it is clear: they have survived but were not evacuated until the second day after the flood. When the first floor was completely flooded, they took refuge on the upper floor, where they held out on chairs through the night while flotsam repeatedly banged against the walls of the flooded house. They remained in the now demolished house the following day as well, hoping their son would come. Since there was no possibility of communication, they did not realize that many areas in the region had been badly hit and roads had been washed out, so that there was no possibility at all of driving to the destroyed areas.

It was only about a week after the flood that Ulrich Schulz was able to visit his parents with his wife and a friend to help them. For a long time, his parents' house was uninhabitable. Not only did the water cause damage, heating oil leaked out and contaminated everything. Due to the forced migration and the destruction in the village, many people were torn out of their social fabric; closely interwoven communities suddenly lacked common places to meet, and some residents moved away for good.

The trauma of the night of the flood and the subsequent displacement left deep traces in people's psyches. Some families had to separate because they could not find accommodation where they could have stayed together. Some seniors who were dependent on assistance went directly to retirement homes because they could not wait for their apartments and houses to be repaired. While it is not documented in numbers, there are reports that many elderly people died in the months following the flood. The uprooting may have adversely affected their physical and mental health. Particularly for families who lived with several generations in the flooding, the scale of the disaster is a rending test. Affordable housing in the region has since become scarce.

Reconstruction in the Ahr Valley

"It's a new start that we didn't choose," Mrs. Petry sums up the reconstruction. Little by little, helpers arrived in the disaster zone. People from all over Germany and even from abroad came to help. And yet, progress was slow. In the Ahr Valley alone, around 9000 buildings were damaged by the flood, and [189] gas and electricity lines were destroyed. For a long time, the Petry/Schulz couple also had to make do without electricity. After some time, they received an emergency generator, with which they could at least operate the refrigerator. But there were also bureaucratic hurdles. While the government aid for household goods and the promised emergency aid were handed out quickly and without complications, the allocation of funds from the reconstruction fund for buildings was extremely slow. Long forms, changing regulations, and complicated procedures overwhelm those affected. All applications had to be filled out online, but many lost their computers during the flood and had no working Internet connection for extended periods of time. Some elderly people faced problems filling in the digital forms to begin with. To get help, it was crucial to take the initiative. But people who were traumatized or have little experience with formal application procedures often did not know where to start. Others fell through the cracks of bureaucracy because the nature of their losses is not covered, or insurance sums only compensate for part of their losses. It also became clear that the cleanup work, which was carried out relatively quickly by the helpers who had traveled to the area, was one thing, but the actual repair work turned out to be far more complex. Craftsmen like tillers, heating installers, and painters, many of whom were themselves affected by the disaster, were needed everywhere at the same time.

Renate Petry tried to cope with this rupture in her life by also helping other people. "For me, it was important that this catastrophe is not just my catastrophe, but that I also use my skills for others." Specifically, that meant taking a closer look where many looked away, taking people by the hand who need help but cannot make the first step on their own out of grief and despair, or are embarrassed to ask for help because the damage to their homes is less severe than others. The disaster brought out the best in many people: generous helpfulness, cohesion, and compassion. But social problems, such as inequality or feelings of envy toward those with insurance, who were able to return more quickly or received more help, also came to light. Support among neighbors is limited by the fact that many are themselves affected and fighting their own battles. All of this must be managed. The guest house "Hinter der Insel" in Üxheim, however, was reopened after the worst of the damage had been repaired and the wounds of the disaster started healing.

Besides basic services, such as train connections, that were disrupted, the cultural landscape and social life of inhabitants had also collapsed. The massive amount of time and effort required for reconstruction in the communities now also meant that for months, there were hardly any other topics of conversation among friends and neighbors. The expectation that everything will be fine again as soon as a house is habitable also created enormous pressure among people whose entire lives had changed since the flood. Especially in places where practically everything was destroyed, there is a collective trauma.

People who had previously stood firm in life have been uprooted by the disaster. The restoration of social cohesion often takes much longer than alleviating economic damages. A return to the status quo ante is often not possible. Thus, the Ahr Valley flood is emblematic of the climate catastrophe as such, which forces us to say goodbye to the world climate in which our grandparents still lived [190]. At the same time, it stands for the missed opportunities at the crossroads of climate protection. In North Rhine-Westphalia, for example, renewable energy projects were massively slowed down by absurd regulations. Very high minimum distances between wind turbines and residential areas prevented the rollout of onshore wind, and in the solar sector, panels were not allowed to be installed on (unused) agricultural land. There was no obligation to install solar systems on new buildings, as in other German states. Moving away from innovations means fueling the climate risks that ultimately everyone must bear. And yet: out of this insight and out of the grief over what was lost, something new can grow. The course can be set anew: for a sustainable, resilient infrastructure and economy. "Out of loss, there is an opportunity to forge new paths. But you can't expect that from severely affected individuals. It takes outside forces to give courage and impetus without immediately imposing a prefabricated plan," explains Renate Petry. The night of the flood in the Ahr Valley is not just a standalone disaster. It is part of a larger trend, the climatic unleashing of natural forces that will claim many more victims. Last but not least, lessons should be learned from the events in the Ahr Valley—for the region itself, but also beyond.

For Renate Petry, one thing is certain: "Money alone will not save us. What is needed is an honest reappraisal of what has happened—to be prepared for the next catastrophes that are bound to come. And then there may not be as much money to repair damage." A change in thinking has already begun among many people; now, it must be followed by action. Mrs. Petry fears, however, that *one* tragedy may not be enough to turn the tide.

Unheeded Warnings

The story of the Petry/Schulz family is one of many fates after the flood disaster in the Ahr Valley that still reverberates today. Not all people took advantage of the time before the disaster to protect themselves as best they could; many had never experienced flooding before and were caught off guard. How could the flood have caught so many people unprepared in the first place? Various institutions, such as the German Weather Service, the European Flood Awareness System (EFAS), and the flood control centers of the federal states, such as the State Office for the Environment of Rhineland-Palatinate, had indeed predicted extreme precipitation and flooding. The EFAS, for example, signaled the highest warning level for the Ahr 2 days earlier.

The German Weather Service also reported extreme rain with the highest warning level on July 13. But the warning chain is long. Whether and when to evacuate, for example, is also up to the municipalities, which often have only limited technical capacities to interpret meteorological data on an ad hoc basis. In addition, rainfall can only be predicted for larger areas. Exactly where life-threatening flooding will occur is more difficult to anticipate. For example, massive rainfall was also announced for the Black Forest area for the same period, without serious flooding occurring there. In this respect, a balance must be struck between a large-scale evacuation and the acceptance of substantial risks. At which point the wrong decisions were made in the flood disaster at the Ahr river is difficult to gauge—in retrospect, it is easy to judge what measures should have been taken. Drawing the right conclusions in the midst of a crisis, on the other hand, is far more complicated and requires, not least, education and training in crisis management. Several committees of inquiry tried to clarify what mistakes were made and how disaster management could be improved. In 2024, several severe charges against officials regarding the Ahr Valley disaster were dropped.

Regardless of the responsibility of individuals, it is clear that the Eifel and Ahr Valley, just like quite a few other regions in Germany, were not prepared for such a disaster. The rainfall was so heavy that people were powerless against the forces of nature. In many places, rainfall amounts of more than 100 L per square meter were recorded within 24 h [191]. In addition, in some areas the enormous rainfall was concentrated in just a few hours—hitting record values. In the end, the masses of water even caused dams to overflow.

As a result, many people decided to evacuate in a very short time and were unable to secure even their most important documents. When the floods reached the residential buildings, rooms filled up to the ceiling within minutes. Some residents even had to take refuge on their roofs. Without the initiative and instinctive reactions of those affected, there might have been even more deaths. Weeks and months of displacement followed. People had to hold out in emergency shelters or found refuge in other communities, with friends or relatives.

Whether displaced people are able or willing to return to the devastated areas depends on many factors. Older people, while deeply rooted in the affected towns and villages, may not have the capacity to rebuild destroyed housing. Others suffered severe trauma and are seeking new beginnings in other parts of the republic. And in certain areas, the degree of vulnerability is so high that rebuilding should be avoided. But in summer of 2023, it became clear that the vast majority of the demolished houses was being rebuilt in the same risky places, mostly for bureaucratic and financial reasons—many insurances only pay if houses are reconstructed where they originally stood. The example of the Ahr Valley flood also shows that few people consider emigrating to another state as long as there is the option of staying in the surrounding area or in their own country. Germany, through insurance, government aid, the willingness and ability of the population to donate, and a solid infrastructure, offers a number of options for coping with extreme weather events that do not exist in most developing countries. Although the suffering due to loss of home is the same, the survivors of the Ahr Valley flood have more doors open to them than storm nomads from Bangladesh or the Philippines. But Germany's prosperity is at risk from the consequences of climate change, not least the horrendous economic costs of the 2021 flood disaster. For example, the federal and state governments have pledged to provide 30 billion euros to help rebuild the flooded areas, which are mainly located in Rhineland-Palatinate and North Rhine-Westphalia. Even before that, dry summers required hundreds of millions of euros in government aid to support forestry and agriculture.

Only very few countries are able to spend such sums on damage events. By way of comparison, the industrialized countries decided at the international climate negotiations in Copenhagen in 2009 to jointly provide an annual sum of 100 billion euros for developing countries from 2020 to 2025 so that they can invest in *both* emissions reductions *and* climate adaptation. Mind you, all developing countries together were to receive this sum. However, the industrialized countries together were not able to raise the sum of 100 billion [192] in 2020 and 2021. Only in 2022, the goal was met [193]. A new climate finance goal was agreed in 2024, which foresees an increase in annual public finance to 300 billion by 2035 and additional sums from other sources, such as the private sector. For losses and damages for climate impacts caused by industrialized countries, a separate fund was agreed in Sharm El-Sheikh which was capitalized with an initial 674 million US dollars. While the creation of the fund was a success, the contributions are far too low still.

Fossil Disaster Management

Reconstruction of buildings costs not only money and time of the affected families, but also other resources. This can be extremely emissions-intensive, especially when decisions have to be made under high time pressure because people—understandably—want to return to their homes as soon as possible. Events like the Ahr Valley flood thus also can translate into a setback for ecological efforts since both the reconstruction work and the disaster relief, which rely almost entirely on fossil fuels, leave an enormous CO_2 footprint. If, as in the Ahr flood zone, the power supply collapses, diesel generators are used. For example, the Germany's Federal Agency for Technical Relief reported diesel consumption of 300,000 L for the first few weeks after the Ahr disaster [194]. There is no question about it: When people are acutely threatened, it is morally imperative to expend even very high resources to save lives. Moreover, it can be challenging to transform disaster management. Nevertheless, a change in thinking must also take place in the humanitarian sector, such as disaster relief, so that sustainability aspects are integrated into operations.

A successive phaseout of fossil energies (first decarbonize industry, then agriculture, then development aid, and finally disaster prevention) would exceed the remaining CO_2 budget. Therefore, all sectors must now transform simultaneously. Clearly, a shift to more ecological processes can only be realized in the medium to long term and must not limit the operability of aid agencies. But more than three decades after the first report of the Intergovernmental Panel on Climate Change, the time has come for the first steps toward climate neutrality.

In some flood-damaged communities, plans are already being implemented to move away from the fossil fuel path. Renewable energy systems could also make villages more resilient. For example, the heating in the village of Dernau, Ahrweiler district in Rhineland-Palatinate is to be converted to environmentally friendly systems as part of the reconstruction process. This is not only better for the world's climate but can possibly limit the damage of future floods.

From the Frying Pan into the Fire: Extreme Precipitation and Climate Change

Extreme weather conditions such as the low-pressure area "Bernd," which caused the Ahr Valley flood due to massive precipitation, will increase in frequency and intensity. Already today, one in four rainfall extremes can be attributed to climate change [195]. It also played a role in the tremendous rainfall in the Ahr Valley. Warming to date has increased the probability of occurrence for this weather event by 3–19% [196]. Comparable to the Ahr Valley floods, in October 2024, South East Spain suffered dramatic rainstorms, which severely impacted Valencia. The region was flooded by up to 500 L of rain per square meter a day—the *yearly* average. Shortly after the disaster, over 200 deaths were counted.

But why does it rain more heavily when the Earth warms up? Among other things, this has to do with the fact that the air can absorb more water when it gets warmer. For every degree Celsius increase in temperature, the water vapor content increases by about 7%. This water eventually pours down. In the case of Spain, high water temperatures over the Mediterranean contributed to the flood disaster.

Long-lasting weather conditions place a particular burden on people. After all, a few rainy or hot days can be tolerated, but if the weather conditions persist, heat waves and floods claim lives in the worst cases. More than 70% of Europe is already affected by such persistent weather conditions during the summer months, making extreme events more likely [197].

Attributing a weather extreme to climate change is a complex undertaking that a growing community of scientists is now grappling with. So-called attribution research tries to determine whether global warming has already played a role in a weather event such as a drought or flood. Scientist Dr. Friederike Otto together with fellow researchers is trying to gain information and insights shortly after the occurrence of climate extremes as to whether and to what extent climate change has had an influence on the occurrence of the weather events.

In addition to scientific classification, the aim is to make information available to the public as quickly as possible. This does not correspond to the classic scientific model, where analyses can only be published many months or even years after an event because they had to go through an extensive review process. But years later, the general public's attention and empathy with the victims have worn off or are overshadowed by the next crisis. That is why Otto, who works at Imperial College in London, and her colleagues publish their already vetted methodology behind each assessment on their website for the broader scientific community [198]. Despite all the existing limitations, she manages to provide robust data and draw attention to it: Climate change is not a distant future scenario—we are already in the middle of it.

Displaced by Coal Mines

Despite the realization that weather extremes have already increased due to inaction in climate protection, fossil fuels continue to be extracted and used at breathtaking scales. Anyone who has ever stood in front of an open-pit mine understands the extent of the direct destruction caused by the use of lignite. In Lusatia, black-gray wounds were cut into the surface over distances that stretch to the horizon, presenting man-made moonscapes in the middle of Germany. It is not only in Bangladesh that villages have to make way for the fossil industry. In the Federal Republic of Germany, too, entire communities have disappeared to make room for opencast mining. One example is the village of Immerath in North Rhine-Westphalia, which has been resettled and then demolished since 2006. Cemeteries had to make way for the coal mania, although other technologies for energy generation had long been available by then. In 2018, even the imposing Immerath Cathedral, St. Lambertus, was razed to the ground despite huge protests by local residents and activists.

The list of buildings to be demolished includes other German cultural assets. With the courage of desperation, environmentalists in Garzweiler in the Rhineland are trying to save villages from destruction by opencast lignite mining. One of them was Lützerath, but the protesters fought in vain. The village that had existed since the twelfth century and its architectural monuments from the eighteenth century were finally sacrificed in 2023 for open-pit mining.

Germany, meanwhile, is one of the world's largest exporters of electricity; in 2016, it even topped the list [199]. In view of the advancing climate crisis, it is absolutely incomprehensible that people are being evicted from their property to make way for lignite excavators—even more so in a country where better technologies for energy supply have been available for decades. This wanton destruction of villages where people have lived for centuries will result in further displacement, far from the coal mine, for example, in regions of the world devastated by climate change. For if the last villages in the coalfields disappear, facts will be created for the further use of coal-fired electricity generation.

Hot Times

Due to the global use of fossil fuels, the average temperature in Germany has risen by about 1.6 °C since the end of the nineteenth century, more than the global average. Similarly strong increases have been recorded in Austria and Switzerland. The deviation is mainly due to the fact that the air over land masses warms up faster than over the oceans. As already mentioned, particular dangers—also for human health—lurk in these temperature changes, for example, when we are hit by long-lasting heat waves. This was the case in 2003, when an estimated 70,000 people died in Europe as a result of the heat [200]. In Germany, the consecutive exceptionally warm summers from 2018 to 2020 cost the lives of more than 19,000 people [201].

Especially in countries where people are not used to high temperatures, adaptation to heat extremes is not yet sophisticated. For one thing, there is insufficient air-conditioning in institutions such as hospitals or nursing homes, and for another, behavioral adaptation is often low. Simple changes in habits, such as drinking more and avoiding the midday heat, are crucial, effective measures, especially during comparatively moderate heat waves. Here, Central European countries can still learn a lot from Spain, Portugal, or India. Because cities heat up to a particularly high degree, architects and urban planners are called upon to turn concrete jungles into resilient urban spaces. Whereas a further increase in temperatures may "only" pose major challenges to people and the environment in some countries, others may be put under existential stress, for example, when already challenging conditions in agriculture are further deteriorating.

A place to get an impression of extreme events in complete safety is the Klimahaus in Bremerhaven, near my hometown of Bremen in Germany. It conveys knowledge about the different climate zones, and it most recently opened an exhibition on extreme weather [202]. It is a place of learning with tangible information about the state of our Earth. There, children, young people, and adults can track down weather phenomena and find out what it feels like to live in a different climate zone. In the energy-efficient building, which is operated CO_2-neutral, scientific findings of meteorology and other disciplines are brought to life. A journey along the eighth degree of longitude, where Bremerhaven is located, with stops in Sardinia, Cameroon, Samoa, and Alaska, among other places, illustrates the way of life of people in different climate zones and the beauty of the planet. Most recently, the Klimahaus opened an immersive extreme weather exhibition, conveying the dangers of climate change impacts and the stories of people who experienced them firsthand.

Threatened Alpine Idyll

The Klimahaus was opened in 2009, and more than 15 years later, the changes that have taken place since are to be gradually documented at all the travel stations and made accessible in exhibitions. In 2021, the first stop to be renewed was Isenthal in Switzerland. For this reason, I accompanied Arne Dunker, the then managing director and co-developer of the Klimahaus, the "traveler," architect, author, and filmmaker Axel Werner who is one of the protagonists of the exhibition, and the writing and documentation team, consisting of travel journalist Anne Steinbach, photographer Manolo Ty, and filmmaker Alessandro Rovere, on their fact-finding journey into the mountains [203].

Seen from the foot of the Alps, it is almost inconceivable that this mighty ecosystem has already been severely disrupted by human influences. Due to ever-increasing temperatures, the glaciers of the European Alps have lost more than 50% of their mass since the beginning of the twentieth century. Their decline is particularly well documented by a comparatively high density of observation data and early photographs. The rate of loss apparently continues to increase, as glaciers melted faster in the 2010s than at any time in the series of measurements. In addition, when permafrost and glaciers melt, entire rock slopes can become unstable. "The ice inside a mountain acts like putty, holding the rock together," write meteorologist Sven Plöger and journalist Rolf Schlenker in their instructive book *The Alps and How They Affect Our Weather* [204].

Glacier retreat is also evident in the canton of Uri, where Isenthal is located. We learn that about 100 years ago, newspapers reported about the noisy calving of the glacier over the mountain edge into Isenthal. Masses of ice crashed daily down the mountainside, and the imposing glacier could be observed from the Biwaldalp. In the meantime, the glacier tongue lies far behind the mountain edge; under the burning sun, this edge retreated further and further into the higher altitudes in the past decades. There, it can still be admired by good mountaineers, to which I do not belong, but more about that later.

The retreat of glaciers has serious consequences. When snow and ice have formed on the mountains during cold winters, the ice melt that begins the following summer feeds rivers and freshwater supplies. However, if the ice is missing, it can lead to water shortages during the hot months. This not only affects agriculture, but also the power supply, because hydropower plants can then no longer be operated at full capacity. Some alpine farms already had their drinking water brought up by cable car, which is labor- and cost-intensive. And in the worst case, it would become the norm: If climate change continues unchecked, the Alps would be almost completely ice-free by the end of the century [205]. This would have consequences beyond the Alpine region because especially in the summer months, rivers such as the Rhine are also partly fed by the Alpine glaciers.

In Isenthal, we stay at the Hotel Urirotstock, where the cordial host provides us with packed lunches for the upcoming hike. In the evening, the hosts play the accordion, discuss, and talk a lot. Then, we set off in the direction of Biwaldalp, which lies at an altitude of just under 1700 m southwest of Isenthal. We scramble up the mountain, take photos, and enjoy the breathtaking mountain scenery and the ringing of the cowbells. Once at the top, we are warmly greeted. Especially the reunion between Axel Werner, who filmed on site more than 15 years ago, and the elderly but still very active alpine farmer Hedy Infanger is heartfelt.

In third generation, the Infanger family manages the alpine pasture every summer with backbreaking work. The cows are driven up the steep mountainside in June as soon as it is warm enough. There is a close bond with the animals. When an animal falls ill, Werner Infanger explains to me, they try to give it the best possible medical treatment. "This isn't just any cow here, this is Josie!" says Werner, pointing to one of the light brown cattle. From the cow's milk, his wife Margrit Infanger, an impressive, resolute person, makes delicious cheese.

At the time of our visit, grass is being cut with scythes on the mountain slopes, raked, and transported away to stockpile feed for the cows. The smell of the grass hangs in the air. In the valley, the controversial leaf blowers with their noxious exhaust fumes, which are harmful to the environment and extremely noisy, are also used for haymaking. At least, they make the hard work a little easier. In addition to the cows, there are also pigs and goats that like to nibble on the flowers in the vase on the wooden benches, and the family takes care of mountaineers who want to climb the impressive Uri Rotstock (2929 m). For this purpose, a couple of sleeping accommodations are available, of which the travel journalist Anne Steinbach and I share one. Anne is a fearless globetrotter who takes her readers to areas that are not listed in standardized travel guides.

While the three sons of the Infangers work on the farm, many young people from the region no longer work in agriculture but have left the villages and only visit occasionally. Although displacement due to climate change is not yet on the cards in this idyllic setting, the risks posed by extreme weather and glacier melt are increasing. Basically, life on an alpine pasture is not without danger. In 2015, a rockfall above the Biwaldalp caused severe damage to the cable car that supplies the alp with essentials. Miraculously, no one was harmed, and the Infanger family home also remained intact. The costly repairs were paid for with insurance payments; otherwise, such an accident could lead a business to the edge of its existence. The risk that the melting of the ice masses will destabilize mountain slopes, causing landslides and rockfalls, is always present. Without government subsidies, alpine agriculture in Switzerland would already be barely economically viable. As climate change progresses, the costs of this agriculture would be driven up even further. Today, it is therefore primarily young people who are leaving Isenthal for other reasons, but it is clear what may happen if rising temperatures and extreme weather events make life increasingly difficult.

Our team wants to film the Blüemlisalpfirn glacier, which has retreated behind the mountain edge. For this purpose, we get up early and start our hike together with the local councilor and former cultural representative of the canton of Uri, Josef Schuler, and the glaciologist Dr. Andreas Linsbauer from the University of Zurich. At first, we hike between flower meadows; then, the terrain becomes more barren. From a distance, we see chamois nimbly jumping over steep rock faces. Further up, we cross two snowfields under the instructions of mountain guide Wisi Infanger [206], who takes good care of us. But shortly before a hut, where we want to take a longer rest, suddenly nothing works for me, and I unexpectedly almost black out. We are at an altitude of just under 2300 m (to explain to everyone who also grew up at sea level: That's actually not high). What looked beautiful before—the mountain panorama, the rock faces—suddenly seems threatening. Josef Schuler sits down next to me and finds a few reassuring words. Everything is spinning, but after a sip of water, I feel a little better. We discuss with Wisi how to proceed. Once again, I try the ascent, but feel more insecure with every step. Josef offers to climb back down with me, and I realize that I might jeopardize the whole endeavor of the shoot if I get serious health problems further up. As much as I would have liked to see the glacier with my own eyes, I have to accept that it is just not possible that day. We descend and, frustrated and exhausted, I sit down in the sun, looking up into the sky.

Maybe it was the altitude, maybe I was not sufficiently trained after a year of corona-owed home office and keyboard acrobatics, or maybe it was just a bad day for me—the mountain brings me to my knees. Ironically, I made it about as far as where the glacier used to stick out its tongue, but now there is nothing but rocks. When Josef and I arrive back at Biwaldalp, Margrit Infanger is playing the traditional alphorn, which is very difficult to learn, and we have a coffee together. A little later, the successful mountaineers arrive. Curiously, I ask them about their experiences on the glacier. They are shocked by the state of the former natural wonder, which has changed so much. Instead of glistening blue-white ice, the glacier tongue of the Blüemlisalpfirn is colored gray-brown by the melting process and the high-frequency debris movements. Though I had been very much looking forward to seeing the glacier, I think in retrospect that my involuntary abandonment of the tour may have had some good. Perhaps, the image of the graying giant, a warning symbol of our destruction of nature, would have been too distressing after all. But perhaps, I will have to return to Isenthal to try once again to take a last look at the glacier, or what is left of it—and to see the Infangers and Josef again in their fascinating habitat, the Swiss Alps, a threatened wonderland worth protecting.

German Coastlines Under Pressure

Glacier melt, the loss of continental ice sheets at the poles, and, in particular, the thermal expansion of the ocean are driving sea level rise worldwide. While Germany, like other industrialized countries, has contributed disproportionately to climate change, it is not as affected by it as other countries. Nevertheless, there are also areas in this country where climate change could force the population to relocate in the future, such as the flat-lying Halligen islands in Schleswig-Holstein. The islands, which have been inhabited since the Middle Ages, are in danger of being permanently flooded by rising sea levels. And there are other problems, too. Increasingly hot summers and torrential rains are also causing problems for the Hallig inhabitants because if their cattle cannot find enough fodder on the meadows, farms have to buy hay from the mainland, which incurs enormous costs [207]. In 2022, sheep and other animals on Hallig Süderoog had to be fed in early summer for the fourth year in a row because the salt marshes were no longer lush green but brown wasteland. The Halligen are an integral part of the cultural landscape of Schleswig-Holstein and popular destinations for tourists, and so the inhabitants hope that they will not be forced to give up their homeland.

Other German islands are also threatened, such as Sylt or Langeoog. In 2022, storm "Zeynep" swept across the North Sea islands and caused massive damage on Langeoog. On Wangerooge, storm "Zeynep" removed as much as 90% of the main beach, sucking it into the sea. Tens of thousands of cubic meters of sand disappeared from one day to the next. The coastal stretches now show meter-high break-off edges. Recovering new sand is expensive. More and more money has to be spent on pumping it up elsewhere and then transporting it to damaged beaches by ship. Sea level rise and storm surges could cause many islands to change their coastlines and, in the long term, their position. This dynamic development is to be counteracted by infrastructure measures such as the construction of dikes and the filling of sand so that people do not lose their homes and businesses. Whether this will ultimately succeed is questionable—there are no unlimited financial and technological resources for these countermeasures.

The more often violent storm surges occur, the more difficult it will be to mitigate or repair the damage. Unlimited climate change would result in the Wadden Sea losing 75% of its area by the end of the century, simply because it would then be below sea level [208]. This would destroy the habitat of many species that need resting and nesting places on sandbanks and in the tidal flats, to whose finely balanced ecosystem they have adapted in perfect form.

The rise in sea level will put coastal protection to the test. If emissions are not quickly reduced, life will be considerably more difficult on the low-lying islands by 2050 and may even be called into question on the Halligs altogether. It is foreseeable that even in a moderate scenario, massive damage will occur by the end of the century. Although the Halligen also "grow" through increased sedimentation, this development will not be able to keep pace with the rate of sea level rise.

Near and Far Effects of Climate Change

Climate impacts do not unfold their destructive power in spatial isolation. They have this in common with pandemics or wars, as the last few years have sadly demonstrated. Faraway shocks can impact our domestic security and threaten our economy. Supply chain disruptions shake up production processes, as has already happened during the COVID-19 pandemic. The 400-m-long freighter Ever Given, stuck in the Suez Canal after a sandstorm, blocked passage through the waterway for 6 days in March 2021, causing enormous damage to industries that depended on timely deliveries. Hundreds of millions of dollars were stranded along with the container ship.

A diverse range of food products also depends on supplies from other countries. Severe, prolonged droughts in grain-exporting countries not only affect the producing countries themselves, but also jeopardize supplies to many other countries. In 2010, for example, an extremely hot summer in Russia triggered a reduction in grain yields, whereupon the Kremlin imposed an export ban. This shortage caused grain prices to skyrocket globally and induced hunger crises. Between June 2010 and 2011, grain prices had nearly doubled. The so-called Arab Spring began during this period, with initial protests also triggered by staple food shortages [209]. Since 2022, the increased prices for food and fossil fuels in the wake of Russia's war of aggression on Ukraine led to high inflation, less economic growth, and political clashes in the European Union. Wars of course also contribute to climate change because it not only destroys the local environment for years, but also causes CO_2 emissions on a large scale. For example, by the movement of military forces or the attacks on fuel reserves or gas pipeline infrastructures, massive amounts of greenhouse gases are released into the atmosphere.

People living in poverty suffer most from the price pressure of food and energy; they lose their savings and not infrequently their livelihoods. But availability and price are not the only factors that determine the outbreak of famine. Socioeconomic inequalities determine distribution patterns, and access to food is sometimes used as a tool of war. While there is abundance in some parts of the world, distribution conflicts over survival rations arise in others. Very often, hunger drives internal migration. Some people, occasionally supported by their communities, try to migrate across national borders, such as to Europe. Thus, money may be collected in the village for an individual to migrate (usually the best educated young men). They hope that remittances from afar will bring funds back to the home village. If the migrants are unsuccessful, shame marks their unfortunate return.

Climate Migration to Europe?

In addition to the economic impacts from distant storms, floods, or droughts, climate change may also make more people want and possibly need to migrate across borders. It is difficult to prove that a climate signal is already behind some of the migration movements to Europe. Because of existing asylum laws, hardly any refugee would cite hunger and poverty as an official reason for their flight, let alone attribute it to climate change, as this could disqualify them for political asylum. Yet, individual testimonies of ever-dwindling resources in source countries suggest that climate impacts are creating ever-greater lack of opportunities. The often desperate situations are frequently further aggravated by a lack of governance, corruption, and, in some cases, systemic violence. As a result, people are trying to build a life elsewhere. However, if their journey leads to Europe, this does not necessarily translate into a better future.

The appalling extent of the suffering in European detention camps for asylum seekers was documented by a group from the German Climate Foundation, which traveled to Greece in 2020 and witnessed the inhumane hardship of the EU border policy [210]. The research was part of an exhibition on climate migration and meant to investigate whether people coming to Europe also left their homes due to climate impacts. Photographer Manolo Ty, video journalist Larissa Rausch, and staff of the German Climate Foundation accompanied by a translator were drawn to Samos, more precisely to the refugee camp Vathy. The group chose this camp because it was less in the public eye than, say, the horror-stricken Moria refugee camp. However, the conditions the team found in Vathy were beyond anything they could have imagined as acceptable on European soil.

The core area of the camp was located on an old military site on a slope above the city. and originally intended for about 650 people. However, because Samos was one of five main destination islands where refugees coming to Greece were to apply for asylum, the occupancy swelled at times to 8000 people [211]. Around the formal camp area, the "jungle" formed, an informal tent city in which people from all over the world lived under the most adverse conditions: young, old, men, women, and children from Syria, Afghanistan, Haiti, Burkina Faso, Nigeria, and many other countries. This informal part of the camp was much larger than the core camp on the military site. In order not to get into the military exclusion zone, the team of the German Climate Foundation stayed exclusively in the so-called jungle and filmed interviews with refugees there, even if the Greek police would later claim otherwise.

In the narrow alleys next to the slum shelters made of plastic bags and tarpaulins, garbage piled up. The packaged food that had been distributed to the refugees was already expired and moldy. Those who did not want to suffer from diarrhea regularly threw it away, except for a few patties of bread. The leftovers piled up on the roadside, Manolo and Larissa tell me. The company responsible for the food supply had already had to pay a fine for their inhumane actions, but the business with the rotten food was apparently still so profitable that fines were hardly a deterrent.

The infrastructure was also far from sufficient for the number of people who crowded the site. Women dug holes in the ground in order to be able to defecate at night. At first, Manolo could hardly believe the descriptions of the refugees, who reported an extreme plague of rats. They said smaller children rarely got to sleep at night because the rats gnawed on them. That is why the little ones often dozed during the day. Then, after dark, Manolo visited the camp again himself: "They were everywhere. I've never seen so many rats. They even ran up on me, jumped on my legs. It was indescribable."

For the refugees, there was no escape from the rat-infested slum. "From 6 p.m. on, they were no longer allowed to stay in the village but had to return to the camp. Some supermarkets did not serve them at all, others had introduced a two-queue system at the checkouts: one queue for refugees and one queue for local residents and tourists," Larissa says. It was a system reminiscent of apartheid. Frequently, internees were also prevented from leaving the camp during the day. Some described bribing the police officers to at least buy something to eat in the village.

As soon as the refugees have submitted their asylum application, they receive temporary cash cards to which an amount of 70 euros is transferred once a month. Because the basic supply is not secured, this amount is not nearly enough. The internees need money for everything: wood for cooking and construction, clothes, tarpaulins, and groceries. For each cash withdrawal, the refugees have to pay a fee of several euros, which the providing bank collects.

Hospitals in the surrounding area have quotas for refugees. If the permitted number is exceeded, even acutely ill people are turned away, according to the statements of several people affected. The coronavirus pandemic also was used as a cover to further worsen the situation of refugees. People who arrived during the pandemic were put in an old prison cell in the police station to isolate them. Women with children spent weeks there before a human rights lawyer from Athens, Dimitris Choulis, got them out. Arbitrarily, refugees were placed in solitary confinement on the grounds that they had symptoms of coronavirus. They were treated this way without testing or any other evidence—mostly as punishment for allegedly insubordinate behavior.

The asylum decision-making process is also opaque. Those affected are not told when they can expect a decision on their future. Some of those the team met had only recently arrived, while others had been holding out for 3 years and were still waiting for the first hearing on their case. "Some had already been formally granted asylum but were not allowed to leave the island of Samos. They were beaten with sticks when they tried to get on the ferry that was leaving Samos," Manolo recounts the reports of NGO workers and interned migrants.

At the same time, there were people who had already received a deportation notice, as well as those who wanted to voluntarily leave the EU again and return to their home country but were prevented from doing so. "Many told us: 'It's hell here. We'd rather go back to war.' But these people were not allowed to return either." The suspicion cannot be dismissed that the overcrowding of the camp structures and the inhumane treatment of refugees are part of the Greek-European border and asylum policy: harshest deterrence of people who have already been taken to task by hunger, war, and dangerous migration routes. According to witnesses, the hopelessness in the camp was so great that there were repeated suicides. Organizations such as Doctors Without Borders or the local aid organization Samos Volunteers tried their best to make the circumstances in the camp more bearable, but their capacities were regularly exceeded by the large number of people in need.

Greek residents who allowed refugees to camp on their land were subjected to great stress. Manolo and Larissa spoke with a pensioner on Samos who had allowed refugees to live on his agricultural land. In order to keep warm during the cold nights, they gradually cut down his 200-year-old olive trees. Now, the pensioner lacks the income from his small olive harvest, but neither the EU nor the Greek authorities pay him compensation.

In the migrant's precarious situation, which was characterized by the fear of being denied asylum or being arrested without reason, it proved to be extremely difficult for the team of the German Climate Foundation to find interview partners who wanted to talk about the causes of flight. Also, there had allegedly already been isolated incidents of unknown persons visiting the camp under the pretext of media interviews and then transmitting the statements of the refugees to the Greek asylum authority so that it could reject more applications. Despite the mistrust, the group around the German Climate Foundation managed to get into conversation with refugees and actually met people whose flight story was related to climatic extremes.

For example, the group spoke with a young woman from the West African country of Burkina Faso who had lost everything due to a flood. Her parents had died, and the family's belongings were irretrievably lost. She had fled to Europe in order to escape her hopeless situation and to complete an apprenticeship in handicrafts. She then wanted to return for a new start in her own country. For her, it was not the big dream to emigrate to Europe, but the last attempt to free herself from her hopelessness and to give her own life a chance. A man from Haiti, on condition of anonymity, told the journalists that tropical cyclones had destroyed fields; thus, the basis of many livelihoods in the country had collapsed. Every time families had rebuilt their homes, they were hit by another disaster, a major earthquake, hurricanes, or extreme rainfall which pushed them back deep into poverty. Hence, climate change was just one of many factors and recurring shocks in a country marked by weak statehood, the Haitian reported. As more and more people around him felt the same way and public order became more fragile, he was no longer safe. He received death threats and could not expect adequate protection from the police. His body was covered with scars. He embarked on an adverse and long flight that eventually led him to Samos via detours.

On the third day of their research, the team filmed on a beach just outside Vathy, where also boats from Turkey cross. Afterward, they got into their rental car. Only a short time later, they were stopped by the police allegedly as part of a general traffic control. The police officers asked the four to accompany them to the police station to check their personal data. Initially, the Syrian-born translator working for the German Climate Foundation, who has a work and residence permit in Germany, was accused of being in the European Union illegally. While this accusation seems completely absurd to an employee of a German foundation, it is extremely dangerous against the backdrop of what is happening in Greece. For example, the case of a Syrian refugee became known who had been granted a right of residence in Germany but was picked up by the police in Greece while he was searching for his missing little brother. The police confiscated his passport and all documents and finally forced the man onto a rubber dinghy, with which he was illegally taken to Turkey. Only after a 3-year odyssey of legal failure, he was able to reenter Germany [212]. Knowing this, the threatening gestures and accusations of the Greek police officers were to be taken seriously.

But that was not all; the group was split up and interrogated separately. They were accused of espionage and threatened with a 25-year prison sentence. An additional accusation was that they had allegedly filmed a restricted military area without permission. Yet the team, which had registered with the Greek authorities prior to its trip, had not even visited the military compound, but had limited its research to the informal settlements outside any military areas.

Harassment was also among the interrogation methods: A journalist's facial mask was thrown away without replacement, in violation of the then valid coronavirus rules; the members of the group were forced to strip naked and were expected to sign Greek language protocols that they could not read. The material that the team carried with them was sifted through, and some of it was confiscated. However, the journalists had protected their sources well so as not to put asylum seekers in danger. After about 5 h, they managed to secretly send an emergency call by cell phone to the German Honorary Consul and a nongovernmental organization on Samos, which mobilized human rights lawyer Dimitris Choulis. Amnesty International and other organizations also became active and began to contact the police station.

Under the mounting pressure, the group was eventually released. There was no official accusation at any time. The treatment of the climate team obviously serves as a brutal deterrent. Journalists are intimidated and threatened to prevent them from reporting on the camp and the conditions on the ground. The treatment of EU citizens raises fears about the treatment of refugees who have fewer potential advocates.

After the German Climate Foundation team left Greece, a tsunami hit Samos and flooded the village of Vathy. Briefly, the situation reversed, and the mountain on which the camp was located became a refuge for the local Greeks. The refugees helped to repair the damage in the village. A few weeks later, the camp, whose formal part was already facing its closure, partially went up in flames. Everything that the residents had painstakingly gathered disappeared in the fire. Now, at an even greater distance from the village of Vathy, a fully supervised high-security camp has been built, which resembles a prison. Although the hygienic conditions here may be better than in the so-called jungle, the deprivation of freedom and surveillance of refugee children and adults are inhumane and have already been criticized by numerous nongovernmental organizations. The camp is monitored by almost a hundred cameras, fenced in by barbed wire, and journalists and lawyers are only allowed access under the supervision of Greek authorities.

As an observer, one gets the impression that the treatment of refugees in the Greek camps is systematically aimed at humiliation, not giving migrants a chance for a dignified life, intent on making them realize that they have no future in Europe. The two-faced nature of European asylum policy is obvious, as Ukrainian refugees arriving since 2022 have received far greater solidarity than other refugee groups. Is this a result of lessons learnt from the mistakes of past migration "crises"? Probably not, as the camp structures in Greece continue to be maintained, and gradually, camps are converted into high-security prisons. One merciful interpretation of the rekindled welcome culture is that EU states feel greater responsibility toward people from Ukraine than toward refugees from other countries. However, the suspicion of racism as a reason for non-equal treatment of BIPOC (Black, Indigenous, People of Color) refugees inevitably suggests itself. The previous recognition of Ukrainian professional qualifications, freedom of movement, and their simplified access to the labor market should be extended to other groups of refugees.

In view of further increasing drivers of migration from the portfolio of climate impacts, the question must also be asked: How should the European Union deal in the future with refugees who migrate not because of wars or persecution, but because of increasingly severe climate impacts? As major emitters, many European states bear a responsibility for climate damage. Without early precautions to protect the livelihoods and the whereabouts of climate migrants, the already foreseeable effects of global warming threaten to put European values to a severe test.

Thesis

Even in Central Europe, people may lose their homes and cultural assets due to climate impacts. Although in global comparison, more options for adaptation are available, rising emissions mean that these opportunities are shrinking and becoming more costly. The cheapest solution remains comprehensive and effective climate protection.

Chapter 8
A Climate Passport for Climate Migrants? The Political Toolbox for the Systemic Crisis

Uninhabitable Lands ♦ Return Impossible? ♦ Prepared for an Emergency ♦ Nansen's Legacy ♦ A Protection Agenda ♦ Climate Migration Mainstreaming ♦ The Nansen Passport for Climate Displaced Persons ♦ Castles in the Air and Walls on the Ground ♦ Debate in the German Bundestag ♦ Voices from the Small Island States ♦ We Will Stay ♦ Inhumane Border Policies

My mentor of many years, the climate scientist Hans Joachim Schellnhuber, put the following climate parable on paper a few years ago: "Imagine that a certain fish species (for example, cod) is driven out of its ancestral marine habitat (say, the North Sea) by the manifold effects of anthropogenic interventions in the climate system, such as heat stress, oxygen and nutrient deficiencies, ocean acidification, and so on. As a result, large schools of this species instinctively try to migrate to alternative areas (such as the Barents Sea) where they have a better chance of survival. Imagine also that these animals are stopped and turned back by the border police at one of the dividing lines between national exclusive economic zones (e.g., the border between Norway and Russia). The reason given by the authorities for turning them back is that 'there is no claim for asylum according to the criteria of the Geneva Convention'. Apart from being impractical, such a scenario seems utterly stupid and cruel. And yet there seem to be many decision-makers who are or would be willing to implement precisely this scenario if the migrants were not fish crossing the Barents Sea but human beings (crossing the Mediterranean on ramshackle ships, for example)."

With these sentences, he describes the paradox that mankind has reached a peak of civilization on the one hand but, on the other hand, draws political borderlines which, in the case of a destabilized climate, can result in a part of our own species being stuck on possibly uninhabitable land. For if left unchecked, massive and rapid environmental changes will inevitably lead to existential problems, especially in countries with little land mass or high population density. Nevertheless, global emissions are rising.

This also applies to Germany, where emissions rose again by 4.5% in 2021 after a brief decline due to the economic slump in the wake of the COVID-19 pandemic. To achieve the German government's self-imposed reduction target of minus 65% by 2030 compared to the 1990 levels, emissions would actually have to *fall by 6%* annually. In a video message in April 2022, U.N. Secretary-General António Guterres called the Intergovernmental Panel on Climate Change's state of the world report from Working Group III, which deals with emissions reductions, "a file of shame, cataloguing the empty pledges that put us firmly on track towards an unliveable world [213]."

Uninhabitable Lands

But when does climate change make an area uninhabitable? There is no universal answer to this question, because both the intensity of the climate impacts and the capacity of societies to respond to them ultimately determine the degree of habitability. Thus, in purely theoretical terms, technological interventions can be used to create habitats even in inhospitable areas, but this involves the expenditure of large amounts of resources and significant funding, which are available only to very few countries. Distributional inequities between and within countries also play a key role in whether populations can adapt. The more extreme the consequences of climate change become, the more difficult potential adaptation measures become. Due to the complexity of these different factors, a number of scientists from several disciplines are addressing the question of the habitability of landscapes under the pressure of climate change [214].

For an area to remain habitable, it must be ensured that the basic needs of the people living there are covered. This is no longer the case, for example, if there is a risk of undersupply, if serious damage to health is to be expected, or if human rights are violated by a lack of access to basic services. The refugee movements in the Central Pacific, Bangladesh, or the Sahel show that migration decisions are always based on an individual risk assessment: How safe is it to stay? What awaits me elsewhere? But when habitats become death zones, people have no choice but to flee. In the course of this century, various areas of the world could lose their habitat function for humans. Our actions today will determine the extent of the changes.

The factors that make an area uninhabitable fall into several categories (Fig. 8.1): The crossing of physiological boundaries, for example, during heat extremes, such as those which may occur in the Amazon or in parts of India in the future (prolonged exposure to the outdoors could be fatal); the loss of land, which is a real risk not only for small island states; the frequency of extreme events, which affect countries

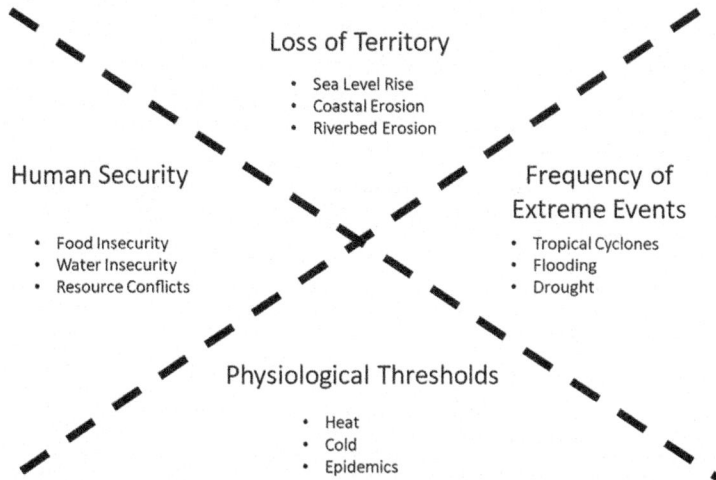

Fig. 8.1 Potential forms of climate-induced uninhabitability

such as the Philippines and lead to famine crises (when harvests fail, agriculture-based livelihoods lose their foundation); and finally, the threat to human security, the interaction of climate change, livelihood loss, and violent conflict, as seen in the Sahel.

If this complex hazard potential continues to grow over the course of the century due to ongoing warming, migration offers the only and last chance to overcome the limits of adaptation. The issue does not only pertain to the question whether an area becomes uninhabitable for all the people living there, but also whether degraded ecosystems can support a growing population, for example. Climatically driven import dependence on staple foods also plays a role in this context. The greater it is, the more likely parallel crises, such as wars, can trigger famines—which in turn may cause migration.

Return Impossible?

If at all possible, most people would prefer to stay where they are at home or return there as soon as circumstances allow. But this option is often denied to them. The United Nations refugee agency UNHCR has already recorded a sharp decline in the number of safe returns of refugees who have fled their home countries for a variety of reasons [215]. While there were 15 million returnees between 1990 and 1999, there were only 10 million between 2000 and 2009 and only 3.9 million between 2010 and 2019. Although this trend is likely to be closely linked to ongoing conflicts and political fragility, serious environmental degradation may also be a factor that should not be underestimated. Organizations such as UNHCR and host countries themselves often urge that refugees be allowed to return once conflicts in their country of origin have been resolved. However, deteriorating environmental conditions, particularly in states that have been rocked by violent unrest at the same time, make the possibility of return even more difficult. For humanitarian reasons alone, it would be urgent to give people who are denied this opportunity in the long term the chance to move on and gain a foothold in another country. For too many people, enduring the uncertainty of a camp has already become a permanent condition, because countries continue to refuse to take in refugees or deny them integration into society.

Prepared for Emergencies

Research suggests that even at the current warming of just under 1.2 °C, climate impacts are destroying the livelihoods of many small farmers and fishermen, forcing them to migrate. The closer global warming approaches a 2 °C scenario, the more frequent climatic changes will occur, causing population shifts of humans and animals. Unchecked climate change, on the other hand, will inevitably lead to massive migration. That is why it is necessary to reduce greenhouse gas emissions worldwide and comply with the requirements of the Paris Climate Agreement.

However, as global emissions have risen again after the brief pandemic-related pause, a worst-case scenario cannot be ruled out. In May 2022, the World Meteorological Organization (WMO) reported that there is already a 50% probability that the 1.5 °C temperature limit will be reached in the next 5 years [216]. During this period, a new heat record will also be recorded with over 90% probability. The projections have become true; in July 2024, a daily average temperature of 17.16 °C was recorded globally, marking a new record [217]. Temporarily, the 1.5 °C threshold has already been crossed, during the El Niño years of 2023–2024. This trend has been accompanied by more extreme weather conditions overall. Thus, unlike in the case of violent conflicts, which can suddenly erupt without years of prediction, climate change as a driver of migration can be anticipated, at least in broad terms.

While strategically planned migration often leads to positive economic effects for migrants and the receiving and sending communities, ad hoc migration as the last possible response to climate change is often detrimental to the affected and receiving communities. To avoid suffering-driven survival migration, political action is needed to support and legitimize migration as an adaptation strategy. National and international institutions, which have not been able to cope adequately with previous migration crises, will not be able to cope with more complex migratory movements due to climate change without an extended mandate and greater funding. There are hardly any practicable legal guidelines for the emergency case when national borders are crossed, for example, because the entire national territory is endangered by sea level rise or has already become uninhabitable (Chap. 2).

However, although there is a significant risk that this is precisely what will occur, and although several institutions emphasize the need to address this issue, few governments have yet developed strategies to deal with it. Some states, such as Fiji and Vanuatu, issued guidelines for internal resettlement (Chap. 3), but the legal gap for transboundary climate-induced migration and displacement remains. The reasons for this are many. In potential host countries and high-migration regions, anti-immigrant rhetoric in many cases prevents public support for legal immigration. Conversely, governments that are severely affected by climate change are reluctant to take a leadership role in regulated migration policy-making, as this could be interpreted as abandoning their country because they believe that the 1.5 °C temperature limit can no longer be met.

In light of the emerging climate impacts, it would be foolish to wait for even larger migration movements to emerge before implementing protective measures. Therefore, the international community should proactively develop instruments that improve the protection of people particularly affected by climate change.

Nansen's Legacy

With this mandate, the so-called Nansen Initiative was launched in 2012 by the governments of Norway and Switzerland and supported by the EU Commission and Germany. It was intended to improve the protection of people who had to leave their own country due to natural disasters or the consequences of climate change. The aim of the project was to promote intergovernmental exchange on this topic and foster cooperation. The initiative was named after polar explorer and Nobel Peace Prize winner Fridtjof Nansen (1861–1930). The Norwegian spent much of his life scientifically exploring the least populated northernmost regions of our planet and led pioneering polar expeditions, such as the crossing of the Greenland Ice Sheet in 1888. But he did not limit himself to his research and to sharing his knowledge with humanity. He was also politically active and used his worldwide reputation to benefit people who were on the dark side of life.

In the later years of his life, Nansen worked to strengthen the League of Nations, the predecessor organization of the United Nations. In 1921, the polar explorer was appointed the first High Commissioner for Refugees. He then devoted himself to the precarious situation of refugees who had become stateless during the World War I, especially Russians, but also Armenians and other affected people, who often found themselves without rights in various European countries. These people were to be able to live and work legally in their host countries with the help of the so-called Nansen Pass. The document, which did not amount to an actual passport because it did not entitle people to vote or confer citizenship, was issued by the League of Nations. With persuasion and charisma, Nansen managed to get more than 50 countries to officially recognize it and thus grant protection to the refugees. Around

450,000 people received a Nansen Pass in the period from 1922 to 1942. The Norwegian was awarded the Nobel Peace Prize in 1922 for his humanitarian work. Only 8 years later, he died at the age of 68 of a heart attack following a severe bout of influenza.

One of his sons, Odd Nansen, a trained architect, founded the Nansenhjelpen (Nansen Aid) for refugees and stateless persons in 1936, continuing his father's work. This organization also supported Germans persecuted by the Nazi regime, especially German Jews who wanted to flee to Norway. After the occupation of Norway by the German Wehrmacht and the installation of a fascist puppet government in Oslo, Odd Nansen was imprisoned in 1942, and Nansen Aid was banned. Initially, Odd Nansen was sent to Grini concentration camp near Oslo, from where he was deported to Sachsenhausen concentration camp northeast of Berlin. Secretly, he wrote a diary. He hid the notes until they could be smuggled out of the camp. Years later—Odd Nansen survived the concentration camp—he published his memories in the book *From Day to Day*.

After his liberation, he continued to work for the needs of refugees; he was particularly concerned with the fate of children. In 1946, he became a co-founder of the children's relief organization UNICEF and, as director of Norwegian European Aid, also continued the work of Nansen Aid. Under the new name Norwegian Refugee Council (NRC), the globally recognized organization is still committed to helping people on the run today. Remarkably, Nansen also advocated for the German expellees from Central and Eastern Europe, most of whom were housed in camps under the most adverse conditions before moving to the West. In an interview, Nansen explained: "I know what it means to live in a camp for years. It means a slow demise [218]." He thus firmly confronted criticism of his commitment and compassion for German displaced persons. After World War II, millions of people had become refugees or internally displaced persons, eking out a living in camps under precarious conditions. Odd Nansen was concerned with creating sustainable protection structures for refugees to prevent such crises: "Wouldn't it be worthwhile to set in motion this greatest rescue work in history for ourselves, for the refugees and for the whole world? [219]" This question is more relevant today than ever, given the record numbers of refugees and displaced persons.

A Protection Agenda

Launched in 2012, the Nansen Initiative published a 50+ page protection agenda in 2015 that bundles recommendations for national governments [220]. It distinguishes between three policy objectives: the voluntary, temporary admission of people fleeing natural disasters and climate impacts who are not given protection in their home country; the non-refoulement of people already on foreign territory; and the development of long-term solutions for people who have no prospect of returning. This list of criteria is intended to serve as a basis for states to decide who is entitled to protection. Various mechanisms are used, such as instructing border guards to allow people who have fled in the context of specific extreme events into the country. In addition, the Nansen Initiative calls for more research and more thorough data collection, as well as support for the care of internally displaced persons following weather extremes in countries of origin.

After 2015, the Nansen Initiative was replaced by the Platform on Disaster Displacement, which advocates for multilateral implementation of the protection agenda and provides space for knowledge exchange and cooperation building. This is urgently needed, because while much has happened in the field of research on climate migration in general in recent years [221] and the knowledge base has grown enormously [222], concrete progress in the protection of climate migrants fleeing across borders has been largely absent.

However, a number of institutions are now at least increasingly addressing the issue. For example, in 2020, the United Nations High Commissioner for Refugees (UNHCR) formulated a strategic framework for its work in the area of climate [223], which states, among other things, that one of its goals is to support countries in improving the legal basis for protecting persons in need. This is phrased relatively broadly, in part because the organization relies on the support of member states. Although UNHCR also assists internally displaced persons, the Geneva Refugee Convention, with its narrow definition of refugees (Chap. 2), forms the core mandate of the organization. The issue of climate-induced displacement is not new to them. As early as 2009, António Guterres, in his capacity as UNHCR's High Commissioner for Refugees, warned at the World Climate Summit in Copenhagen: "Climate change could become the main cause of flight [224]." Another ray of hope is that the issue is enshrined in the UN migration pact [225]. In it, 164 states agreed on a text that also recognizes climate change as a driver of migration, discussed in the second chapter of this book.

An early champion for the rights of climate migrants is Koko Warner, a scientist who now heads the Global Data Institute of IOM in Berlin, Germany, and previously led the Climate Impacts, Vulnerability and Risks Unit at the *United Nations Framework Convention on Climate Change* (UNFCCC). She has been advocating for better protection of affected people long before the Paris Climate Agreement and has led a number of research programs that have resulted in a better understanding of the complex issue. Warner thereby forged a pioneering path for both research and practitioners.

A recent initiative introducing positive framings around migration opening adaptation pathways and unlocking economic potential has taken foothold: the Global Centre for Climate Mobility (GCCM). With regional foci in the Pacific, Caribbean, and Africa, the center seeks to enable communities in climate hotspots to increase their resilience. It moreover brings together multi-stakeholder groups from policy, science, and civil society in multilateral fora, such as the COP or the New York Climate Week. In particular, GCCM also engages with youth leaders, connecting them across regions.

Climate Migration Mainstreaming

While there is division on the international level on how to best address climate migration, the political anchoring of the issue is also not without obstacles at the national level. In theory, different ministries could deal with climate migration issues, such as those dealing with the environment, development, foreign affairs, interior, labor, health, or even security. In reality, however, the issue often falls between the dividing lines of the traditional division of ministries, and thus, there are no clear responsibilities. An interdepartmental approach would make sense, but it is difficult to implement because the individual ministries often differ in the way they work.

The American position on the treatment of climate migrants has been changing with the different administrations. While the Trump administration denies climate change and even refused to sign the UN migration pact, former US President Joe Biden issued a decree right at the beginning of his term that provides for the strengthening of resettlement programs for refugees and also addresses the issue of climate migration [226]. To be better prepared for it, an interdepartmental report was even ordered. Many hoped it would provide a concrete program of measures, but the paper does not meet these expectations.

The White House report draws on the existing academic literature and assesses possible foreign, development, and security policy implications from an American perspective. In doing so, the writing team links the foreign and domestic policy dimensions of future displacement: "Inadequate policy frameworks to manage large migration flows may exacerbate resource inequalities, stress public budgets, and contribute to xenophobia that increases political tensions [227]."

In addition, it is pointed out that the (human) costs of migration for migrants themselves as well as for the places from which they come are already enormous and could grow even more in the future. For example, if the brightest minds migrate or if there is a shortage of labor in agriculture, the consequences for sending communities can be dire. Host countries, on the other hand, can reach the limits of their capacity when migration numbers suddenly rise, for example, when it comes to supplying and housing people. This is one of the reasons why long-term solutions are needed instead of ad hoc responses to immigration. If no adequate answers are found to these developments in sending and receiving countries, democratic states could be destabilized in the medium term, according to the White House report.

The report concludes with a number of recommendations, including the suggestion that the lack of protection for climate migrants should be addressed in the US Congress. However, the chances of success for this have shrunk to zero since the re-election of Donald Trump. While his compatriots' homes were destroyed by raging wildfires in California, Trump ordered the exit from the Paris Climate Agreement, cheered on by crowds. Even the report's authors acknowledge this, citing a split between the Democratic and Republican parties as a reason for the precarious situation of many immigrants: "The lack of bipartisan agreement on humane border procedures and immigration polices complicates U.S. efforts to mobilize global support for protecting refugees, asylum seekers, and other vulnerable migrants." Thus, the United States is also in a difficult negotiating position when it comes to persuading other countries to take in larger numbers of refugees or even climate-displaced persons. Even more, the United States has been falling short of its fair share for international climate finance for years, in part because the US Congress has blocked the spending of larger sums. At the 2021 COP in Glasgow, for example, former President Biden announced plans to increase the US contribution to international climate finance to $11.4 billion by 2024, but only a meager $1 billion was appropriated by Congress in 2022. Again, in the spending bill for 2024, there was also only $1 billion for international climate finance appropriated by Congress, though the Biden administration had requested more. US agencies had ramped up their finance through discretionary funding and delivered a total of $5.8 billion in 2022, which signaled general commitment to the pledge made in 2021. With the dismantling of the USAID and the repeated American exit from the Paris Agreement, the US government has denied its responsibilities. Germany, with a much smaller GDP than the United States, provided 5.7 billion euros in 2023 and plans to increase its annual funding to 6 billion euros by 2025. This money is intended to help developing countries operate in a climate-friendly manner and adapt to extreme weather events and other climatic changes.

While some countries refer to the challenges associated with climate-related internal migration in their national adaptation plans, multilateral efforts to protect transnational migrants are still in their infancy. The Commission on the Root Causes of Displacement, established by the German government in 2019, also recommends "to acknowledge the legal and protection gap for climate displaced persons and to develop solutions [228]." However, although numerous documents directly or indirectly address the need for new instruments and mechanisms to protect climate displaced persons, proposals for concrete measures are rather scarce—partly because each of these proposals immediately attracts criticism. Nevertheless: Urgently needed is significantly increased funding for adaptation measures, such as securing agricultural livelihoods to enable people to stay and to create pathways for self-determined migration in dignity. That is why freedom of movement for people from high-risk areas should no longer be blocked.

The Nansen Passport for Climate Displaced Persons

One step in the direction of freedom of movement would be, for example, the issuance of travel documents building on the legacy of the historical Nansen Passport, which enabled stateless persons to live legally in other countries as early as the 1920s [229]. Together with the German Advisory Council on Global Change (WBGU), which brings together an interdisciplinary group of renowned scientists, Hans Joachim Schellnhuber developed the idea of a climate passport. Against the backdrop of past refugee crises, he said, urgent provisions need to be made to provide people with travel documents that will enable them to reach safety in the event of an emergency. "It is almost unbearable to imagine that in the decades to come millions of climate migrants would have to rely on human traffickers like those who are currently causing such terrible human misery in the Mediterranean [230]," says the WBGU. Therefore, the council proposes the introduction of a climate passport: "Following the humanitarian innovation of the Nansen Passport, the climate passport should open up early, free and dignified migration options [...] for people existentially threatened by climate change."

This passport should be recognized by countries that have contributed significantly to global greenhouse gas emissions, i.e., in particular by industrialized countries. Although the establishment of the passport should be legally linked to the United Nations Framework Convention on Climate Change (UNFCCC), it should be possible for a small group of states to implement it first and then push it forward, similar to the Nansen Passport. Such a pioneering group would circumvent the obstacle of having to unite the entire community of states behind the proposal. The WBGU's recommendation is to first create a manageable quota for residents of particularly threatened island states and then to expand the group of potential recipients as needed. The idea is compelling. In this way, a preventive solution would be worked out for a small but undoubtedly highly endangered population group, which could possibly act as a precedent for the protection of further climate displaced persons [231].

The council also calls for the following: "The climate passport should be established and financed not instead of, but alongside climate protection and physical adaptation measures. Affected individuals should be able to decide more freely whether and when to migrate through safe and early migration options." Climate impacts curtail the self-determination of many people. Cultural practices, traditional forms of agriculture, and the right to remain in one's home are all being challenged by climate change. While migration is often described as an adaptation strategy to changing environmental conditions, this formulation obscures the fact that moving from rural to urban areas, or even across national borders, generally is a radical turning point in people's lives. This is because, unlike, say, the construction of a sea wall, migration protects lives but not lifestyles. If people are forced to tear out the roots of their previous lives because of greenhouse gas emissions brought about by distant countries, it should be necessary to restore a minimum of freedom of choice, for example, by means of a climate passport.

The idea of such a passport has since been taken up many times. For one thing, the German Green Party introduced a motion on the subject in the Bundestag in 2019, but more on that later. The climate pass also features in the Berlin Futurium's permanent exhibition on future mobility. The Futurium is an interactive forum on the subject of the future, its challenges and possibilities, where people can vote on developments themselves and experience various new technologies and models of society in an innovative manner. Whether these will prevail is uncertain, but this precisely is the charm of the future.

Castles in the Air and Walls on the Ground

In the midst of a wide variety of crises, we may be a long way from implementing an idea like the climate passport. But perhaps, there is a silent majority that would certainly like to see more humane migration policies and greater support for people affected by climate change. In any case, it is worth fighting for such goals, even if the climate passport may still be a pipe dream at the moment.

That unlikely proposals can gather support is well known in migration deterrence. In his first term US President Donald Trump gave rise to the idea of building a wall between the United States and Mexico, several hundred kilometers long, to prevent Mexican and Central American immigrants from entering the United States. For many people, the inhumane absurdity of this proposal initially provoked only a sneering frown. It was almost inconceivable that Trump could receive approval for such an idea in the United States, a country built on immigration. But despite considerable criticism in liberal media, Trump did not let up and stirred up resentment. The effect he had hoped for occurred, and fear and frustration in parts of the population resulted in demands for ever more rigorous border protection. The building of a wall more than 10 m high began, and families were separated from their children when they reached the crossing. Images from the border region in 2021 of white policemen on horseback hunting black migrants from Haiti with whips are reminiscent of the darkest times in history. Then President Joe Biden too expressed shock at the images. But, being distracted by other crises, the outrage of those who saw migrants as human beings faded. More recently, in 2024, resentment against immigrants was steered by false information about Haitian migrants supposedly eating pets, spread by Trump and his running mate JD Vance. Migrants continue to be used as scapegoats by politicians, often leading to their marginalization and deprivation.

In 2023, the Biden administration resumed the border wall construction, pointing out that the money for the extension of about 20 miles had already been appropriated by Congress. During the presidential race 2024, Democrats pledged tougher action against asylum seekers whose number per month had reached an all-time peak. In June 2024, Biden signed an executive order to close the border with Mexico if illegal crossings surpass an average of 2500 people per day. But in his order, Biden also encouraged legal pathways for immigrants.

So, is there anything to be learned for the climate passport from this cruel episode of US immigration policy? From my point of view, if it is possible to generate political majorities for aberrant, hate- and fear-driven ideas, it should be equally possible, with perseverance, to popularize good ideas for which no majority is initially in favor. Only asking for initiatives which already have obvious majorities, where there is no more convincing to be done, will not suffice to address the climate crisis, which puts everything at stake. Some parts of Trump's wall were badly damaged by extreme rainfall in Arizona in 2021 after prolonged droughts [232]. No wall stands forever.

Debate in the German Bundestag

To better protect climate migrants, what is needed above all is the political will to break new ground and learn from past crises. In 2019, the German Greens, then an opposition party, submitted a motion entitled "Climate-related Migration, Flight and Displacement—A Question of Global Justice [233]." It states that advancing global warming and the existing protection gap toward climate displaced persons in international law have already led to a crisis of justice. In this context, the responsibility of industrialized countries is brought into focus: "As—historically and currently—the main contributors to global warming and as globally influential multipliers, Germany and the European Union are crucially important against this background." The motion formulates several demands in connection with climate-related migration, for example, the strengthening of disaster preparedness in climate change hotspots and the protection of particularly vulnerable and marginalized groups, which in many places include women.

In addition, the German government is called upon to "also promote the introduction of a climate passport nationally, Europe-wide and internationally and, in an initial phase, to offer it to the populations of small island states whose national territories are becoming uninhabitable as a result of climate change." In particular, the youth organization of the Green party, the Green Youth, is making a strong case for this proposal. Climate migration should be addressed at an early stage in order to give migrants room for maneuver, which would be further reduced by the climate crisis.

The motion was hotly debated in the Bundestag in 2019 [234]. The extremist rightwing AfD parliamentary group in particular repeatedly put forward absurd arguments with which it played down the effects of climate change. "Climate refugees" were called "green fake news."

The AfD parliamentarian Markus Frohnmaier went so far as to prophesy that if climate refugees were given freedom of movement, the incidents of the New Year's Eve in Cologne in 2015 would occur "all year round." He was referring to the numerous charges of theft, insult, and, in particular, sexual harassment filed in the aftermath of the New Year's Eve celebrations around Cologne's main train station. Men from North Africa were mainly blamed for the assaults, although German men were also among the alleged perpetrators.

What followed was that right-wing populists skillfully used one progressive issue—women's rights and their protection from sexual assault—to play it off against another progressive issue: the right to asylum. The "welcome culture" tilted in many parts of society. Refugees were met with mistrust, and the incidents were misused as an objection to their admission—as was also the case years later in the debate about the climate passport. However, there were hardly any convictions in connection with New Year's Eve; only three people were found guilty of sexual assault and several dozen of theft or robbery [235].

In the debate in the plenary hall, however, Volkmar Klein, a member of the Bundestag for the CDU/CSU parliamentary group, also called the climate passport an "absurdity [236]." Germany was already a pioneer in climate financing, he explained, and the lack of prospects in developing countries such as Chad was linked to poor governance, population growth, and the overexploitation of resources. The consequences of climate change played only a minor role, if any. He did not mention the fact that per capita resource use in Chad is far lower than in industrialized countries, for example, nor did he mention the situation of the small island states that are at the center of calls for safe migration routes.

Voices from the Small Island States

How do the affected people themselves—people in particularly vulnerable regions—think about the issue of the right to transnational protection in the case of life-threatening climate impacts? In the course of various research projects on island states, colleagues from the Potsdam Institute for Climate Impact Research and I, with the help of researchers from the University of the South Pacific in Fiji and a local advisor from the Philippines, asked some interviewees the question: "Should people who are severely affected by climate change in their country have the right to live and work in another country?"

Speaking out in favor of the right to resettlement or the admission of migrants can quickly attract criticism. In order to obtain answers that were as open as possible, we anonymized the statements—and thus arrived at very reflective and multi-faceted assessments of NGO employees, environmental planners, government representatives, and employees of regional organizations.

In the Caribbean, for example, policymakers are increasingly concerned about the magnitude and frequency of tropical cyclones and their impact on displacement. The 2017 hurricane season was etched in the memories of the entire region (see Chap. 3). Influenced by these experiences, most people I exchanged views with favored cross-border freedom of movement. One reason for this could be existing regional agreements of the OECS (Organization of Eastern Caribbean States) and CARICOM (Caribbean Community), which offer Caribbean member states extensive freedom in (labor) migration.

In the island nation of St. Lucia, on the other hand, an official from the ministry responsible for disaster management supported the possibility of migration but, at the same time, pointed to potential problems: "Yes, you have a right to it. It's everyone's moral responsibility that we can all live a good life. But in practice, it's harder for the government to justify spending money on migrants when its own population is doing poorly." Especially in very poor regions, migrants potentially compete with poor residents for resources. One person working for a Caribbean Ministry of Labor emphasized the motive of solidarity: "At the end of the day, we are all human beings. You can't isolate yourself because you never know when you'll find yourself in a situation where you need help too. You never know when you're going to be in a situation where you're forced to migrate."

We Will Stay

In addition to the generally positive response, there were also reservations about potentially top-down relocation of people, which is immediately associated with the issue of climate migration: "If we are not able to relocate people from a certain area within the country at the local level, how are you going to get people to relocate to another country before something happens, and even after something has happened? [In] Anguilla [...] the drought was so bad that the national government [...] talked with the governments of Trinidad and Guyana to move Anguillans there. But only a few moved. The majority stayed. There is resistance to relocation, people are saying, 'we will stay,'" explained a civil protection worker in Anguilla. Many people prefer not to leave their homes and migrate only when they have no other option left. This is no different in the Caribbean than in North Carolina, where more than 160 people died and many houses destroyed by Hurricane Helene in 2024.

In the Pacific, debates about resettlement and displacement have gained much political and public interest since the government of Kiribati purchased land in Fiji in 2014 to secure people's long-term livelihoods (Chap. 3). One man from Fiji, whose work includes involvement in international climate change negotiations, summarized critical issues, including the question of climate justice and the need for integration into the host society: "As a Fijian, I would say yes [people should be given residency rights], but I also look at it from the perspective of the interests of the country where you want to work. [...] I would say 'yes' if climate change is the determining factor, but how can you prove it's a challenge? The fact is that you can make the moral claim that industrialized countries have caused these problems and that we are suffering because of them. We need to reach out to these industrialized countries. In the developed countries like Australia, New Zealand, the European countries, Canada and America, the walls have gotten thicker and higher, and they're not letting anyone in."

The issue of climate justice and the responsibility of developed countries came up again and again in the conversations. An NGO worker from Vanuatu responded, "Yes, and hopefully the more developed countries will open their doors to those who are severely affected by climate change." Several interviews cited the duty to provide humanitarian assistance as a motive for granting the right to migrate to another country and pointed to the existential threat posed by climate change: "Yes, there should be systems in place to help those affected by climate change to live and work in another country. We all live on one planet, and other countries should help when fellow human beings are in need, especially when it comes to their survival," said a Vanuatu official. However, in both the Caribbean and the Pacific, high-ranking officials and experts often answered the question as if they were personally affected by a necessary resettlement: "People from Vanuatu [...] feel at home in Vanuatu. If you rip them out of their home and put them in a foreign country with foreign people ... it would be difficult. [...] If something were to happen in our place, we would prefer to be relocated to an island inland." Their awareness of climate risks and the caesuras that can result from such disasters is evident.

Concerns were also expressed about whether a targeted migration policy could ultimately lead to human rights violations, for example, if migrants were not granted the same rights as long-time residents. In this case, there would be a danger that migrants would become second-class citizens. In response, a scientist from Fiji, who also works on climate adaptation, pointed out the danger of misusing such a protection instrument: "Yes and no. Yes, if people are affected by climate change, they should be granted some kind of status if they have to migrate. No, because there is a risk that the procedures for migration involved will be abused in another country. Before people affected by climate change seek refuge in another country, there needs to be proper planning and guidelines, and those procedures need to take into account the citizens of the host country." Behind this is the fear that forced relocations could occur for political motives. Fiji itself has shown solidarity by selling land to Kiribati, but here, too, fears are raised that if too many people suddenly arrive, conflicts of interest could arise between the local population and migrants. One interviewee, who works in the private sector in Fiji, emphasized the voluntary nature of humanitarian aid: "You're definitely not entitled to it, it's for humanitarian reasons."

In some of the interviews in the Caribbean, it was emphasized that the decision to migrate or remain should be up to the affected person or household. "They should have a right to do it, but it should be the personal decision of the victims of climate change to which country they want to be resettled," said an NGO worker in Kiribati.

The voices from island states are diverse, but they highlight the need to adapt the international migration regime to the new realities of a changing climate. They reflect the growing need for mechanisms that allow climate migration to be handled as flexibly as possible. These include not only a possible climate passport, but also the issuance of humanitarian visas, the expansion of the non-refoulement principle, and the reform of scandal-ridden immigration and border control agencies, including the European Frontex and the US ICE (United States Immigration and Customs Enforcement).

Inhumane Border Policies

Recent reports on the extent of impending climate impacts by the Intergovernmental Panel on Climate Change (IPCC) and the growing suffering of many people whose livelihoods are being parched, burned, or engulfed by floods make it clear: the human costs of maintaining the existing strict border protection and narrow asylum regime will rise in a changing climate. They are already unbearably high, as the relatives of the children, women, and men who have met their deaths on the Mediterranean testify. The vehement defense of national borders has now become a crucial tool for perpetuating global injustice, and climate change is exacerbating this imbalance.

How refugees can be treated humanely was demonstrated in 2022 by the treatment of Ukrainians displaced by war: Hundreds of thousands of war refugees arrived in European countries after Russia's attack on Ukraine, and they had to be cared for immediately and given immediate temporary protection status without asylum procedures. That this unfortunately cannot be taken for granted was made clear by events at the Belarusian border with Poland, where Afghan and other people in need of protection were turned away just weeks before the war began. Equally shocking was the news that BIPoC individuals, such as exchange students from African countries, were prevented from leaving Ukraine. These dichotomies in European solidarity push us to the questions: To whom do we grant empathy? What responsibilities do we accept? And which obviously justified demands for less CO_2 emissions, less destruction of biodiversity, and more practiced solidarity with victims and heroines of the climate crisis are rejected because they are not compatible with our way of life and self-image?

Thesis

A wealth of tools and recommendations to protect climate displaced people are waiting to be implemented. Drawing on existing knowledge and mustering the political courage for new directions could improve the prospects for many people whose future will be determined by the climate crisis.

Chapter 9
Pathways Out of the Crisis: Fragments of Hope

What Climate Migration Has to Do with Racism ♦ What Climate Migration Has to Do with Sexism ♦ Gender Justice in Industrialized Nations ♦ A Social Problem ♦ Solutions Outside the Box ♦ Architects for the Poorest ♦ Floating Cities as a Lifeline? ♦ The Imperfect City of Tomorrow ♦ Climate Knowledge and Culture ♦ Heat Waves in Spring ♦ What's Next? Judgment, Conviction, and Empathy

Many upheavals on the blue planet are already in full swing—and even gaining speed. The previous chapters on different regions of the world reflected the human face of climate change, a face that shows traces of suffering, despair, and devastation: livelihoods, homes, and farms destroyed by floods; parents whose children perished in superstorms; and farmers searching for work in the urban black markets. In contrast, there is another narrative, one of action and survival. These two aspects are the subject of this concluding chapter: the scale of destruction and inequality and the revolt against it—for solutions and ideas that could be groundbreaking. These are fragments of hope which, in the face of multiple crises, challenge our imagination as to how we can still turn the tide. This requires looking back and looking forward at the same time because only by understanding past failures, appropriate answers to the challenges of the present can be found.

What Climate Migration Has to Do with Racism

The unequal distribution of the cause of climate change, global greenhouse gas emissions, has been one theme of this book. Is this imbalance between developed and developing countries related to racism? Economic anthropologist Jason Hickel calculated the share of historical emissions by individual countries that exceed greenhouse gas emissions consistent with planetary boundaries. The United States and Europe (including the European Union) combined account for a 82% of global excess emissions, while the Global South accounts for only 8%. The remaining 10% is distributed among other countries in the Global North [237].

For scientists Ulrich Brand and Markus Wissen, this one-sided overexploitation of the atmosphere is part of the "imperial way of life" based on exploiting both ecological and human resources in the form of labor for the benefit of a few [238]. Even if one does not have to share all of the author duo's conclusions, their analysis of the mechanisms of our economic system is razor-sharp: "EU policy becomes comprehensible as an attempt to defend a prosperity that also comes at the expense of others against the claims to participation of precisely these others." The isolation of refugees is meant to preserve the exclusivity of the imperial way of life because it only remains stable if it can impose its costs on others, according to Brand's and Wissen's critique.

Activist and lawyer Elizabeth Yeampierre goes back one step further in history and sees the beginnings of slavery as the start of systematic environmental destruction: "With the arrival of slavery comes a repurposing of the land, chopping down of trees, disrupting water systems and other ecological systems that comes with supporting the effort to build a capitalist society and to provide resources for the privileged, using the bodies of black people to facilitate that [239]." So, the oppression of people and nature often goes hand in hand, the activist said. Yeampierre concludes, "climate change is the child of all that destruction [...]."

Even though global upper and middle classes are driving emissions the inequality of distribution cannot be denied and is pointed out by Hickel in his country analysis and Yeampierre, Brand and Wissen in their critique of the imperialist features of the industrial way of life. That BIPoC in the Global South are displaced due to climate impacts mostly caused by white people in the Global North is thus both a product and perpetuation of racist oppression.

But even within industrialized countries, the gap between white people and BIPoC widens, which becomes particularly visible when extreme climatic events tear apart the social fabric. For example, in the United States, African Americans and Latinxs were disproportionately affected by tropical cyclones Katrina (2005) and Harvey (2017) [240]. This is also due to structural racism. Lower educational opportunities and incomes mean that in the United States, many BIPoC live in neighborhoods that are more likely to experience flooding because, for example, rents and property prices are still affordable there. In these areas, there are also fewer green spaces that could absorb water in the event of extreme precipitation.

A report from the US National Academies of Sciences finds that urban flooding kills or injures BIPoC more often than white people [241]. Recovery is also slower because BIPoC are not as well insured on average; this again is most likely related to poverty and fewer educational opportunities. In addition, many BIPoC have less confidence in government agencies, which include those commissioned with reconstruction [242]. Given massive police violence against BIPoC in the United States, distrust of government institutions is not surprising. It does mean, however, that people affected by the storm may not be demanding the help to which they are entitled. In the case of the extensively documented aftermath of Hurricane Katrina, for example, it was shown that BIPoC had to hold out longer in temporary shelters than white people and, on average, were also more likely to be permanently displaced [243]. African American women were particularly affected, according to a study of a New Orleans suburb. They "were least likely to be evacuated with their families, less likely to have families or relatives in their first and second destinations, and remained in their destinations for a longer period," thus had to live longer in displacement [244].

Another example: Neighborhoods that were previously segregated in the United States through redlining are now demonstrably exposed to higher environmental risks than other neighborhoods [245]. Redlining in the United States refers to the racist practice of segregating ethnic minorities by excluding entire neighborhoods from investment, resulting in, among other things, reduced access to city services and lower property values. While this practice is now illegal, the effects are still felt today—especially for ethnic minorities who continue to live in such neighborhoods. These districts are particularly severely affected by heat and are on average 2.6 °C warmer—some neighborhoods are even 7 °C warmer than others [246]. Inadequate urban planning and the paucity of tree cover contribute to redline neighborhoods heating up more. For example, many highways cut through these neighborhoods, contributing not only to warming but also to deteriorating air quality [247].

There are similar phenomena in the Federal Republic of Germany. A number of studies in Germany indicate that a "migration background" (which is statistically recorded in some surveys) is even more decisive than the level of income in determining whether a person feels exposed to environmental risks, such as poor air quality [248]. Moreover, BIPoC are still not given enough voice and attention when it comes to climate change impacts in panel discussions and in the media.

Journalist Alice Hasters highlights another point: the extent to which climate change affects existing inequalities. She explains the existing imbalance by saying that "*white* supremacy and patriarchy are fighting back tooth and nail—and we're running out of time to have these discussions because climate change is fast-tracking toward irreversibility, and with it will come new problems [249]." Existing and future climate impacts will inevitably take up more and more capacity and attention. This will leave less space and time for addressing other crises, such as key societal debates around racism and equal opportunity. In a high-emissions scenario, questions of justice could fade into the background, if only because more and more people would be concerned with meeting basic needs.

In the balance sheet of climate change to date, there is also a loss that is hardly noticeable in the public perception. This is the potential of those who have to fight for their survival every day and therefore cannot devote themselves to other tasks. This potential is lost to human progress as a whole. The power to shape their own future is taken away from them by the structural violence of global emissions. Haster's book *What White People Don't Want to Hear About Racism, but Should Know* holds another insight for its readers, which can be applied both to the racism it addresses and to addressing the climate crisis: "The truth is that an open heart, good will, and enthusiasm alone won't save the world [250]."

What Climate Migration Has to Do with Sexism

Climate change, migration and displacement affect men and women differently. For example, in the 2022 Sixth Assessment Report, the Intergovernmental Panel on Climate Change Working Group II notes that the impacts of climate change have already had a detrimental effect on gender equity [251]: "Women are often disproportionately affected by the negative impacts of extreme climate events. The reasons range from care work to lack of control of household resources to cultural dress norms [252]." In many regions, moreover, it is men who migrate in response to worsening environmental conditions, while women often stay behind or leave at a later date [253]. There are several reasons for this: In order to migrate, women often face higher risks and disadvantages than men, such as in terms of their educational opportunities, lack of access to contraception, and traditional norms—all of which affect both their ability to cope with climate impacts and their migration decisions.

Looking at socioeconomic inequality from a global perspective, the dimension of this crisis becomes even more apparent. A report by the aid organization Oxfam reveals that the 22 richest men in the world own more than the entire female population of the African continent [254]. In this comparison, it is hard to decide what to be more shocked by, the boundless wealth of the global superelite or the abject poverty in which many women are trapped. In general, women often have less wealth than men because care work for children or relatives is not remunerated. Even if working in the fields helps ensure self-sufficiency, small farms rarely yield substantial profits. Without any financial reserves or access to bank accounts, women are exposed to considerable risks. Poverty usually means dependence, insecurity, and deep cuts to a person's basic freedom of choice.

Often, existential hardships, lack of gender equality, and climate impacts coincide. For example, many people suffering from water scarcity live in areas where there is great poverty and inequality between women and men is enormous [255]. Of particular concern is the lack of access many women have to adequate and nutritious food. "The sex gap in access to food increased from 2018 to 2019, with women living in rural settings the most affected—paradoxical since women and girls represent most food producers and food providers," concludes a 2022 United Nations Development Programme report, citing a number of studies [256].

However, it is problematic to see women only as victims of the climate crisis. In many societies, women already occupy key positions in which they prevent resource conflicts, practice sustainable forms of land use, implement adaptation measures to climate change, or even bring about emission reductions. In this respect, women are important actors in addressing the climate crisis. But in order to take action on a large scale, cornerstones of development need to be re-laid. For example, if women are given access to electricity, education, and modern cooking facilities, i.e., if they no longer have to spend hours gathering wood, they will have more time to take part in training and education. It has been shown that higher levels of education also reduce the number of children they have [257]. This, in turn, can reduce future resource needs, and women are empowered by fewer caregiving responsibilities.

In my research, I have repeatedly encountered impressively strong women who are taking their fate into their own hands. In Bangladesh, I spoke with a group of three women who first escaped their violent husbands and then had to migrate because tropical cyclone Aila had destroyed their shelter.

Women are often exposed to particular risks when they have to flee alone to countries where women's rights are poorly protected [258]. Unfortunately, this also affects many countries severely affected by climate change, such as Bangladesh, which ranks 133rd in the Gender Inequality Index, or Burkina Faso, which ranks 147th [259].

The involuntary path into sex work is not uncommon after natural disasters in countries without functioning social security systems. But the women's group with whom I exchanged ideas in Bangladesh took advantage of a "cash for work program" and helped build dikes and was thus able to provide at least the most basic necessities. They resisted a system in which the majority of women are assigned their roles by men.

Gender Equality in Industrialized Countries

Discrimination against women can vary widely and occur on multiple levels. For example, black lesbian women often experience more discrimination than white heterosexual women, and women with disabilities more than those without. That is why an intersectional view is important—which, by the way, includes the fact that women also block the development of other women. This begins with the fact that women in industrialized countries contribute significantly to emissions that deprive women in developing countries of their livelihoods.

However, the lack of equality for women is by no means limited to distant countries. Even in industrialized countries, the distribution of power and resources between men and women is still not balanced. In climate science, for example, there are significant disparities between men and women, which are further exacerbated by intersectional inequality. Reuters published a list of the 1000 most influential academics in climate science in 2021. Among the 1000 listed, only 122 were women, and only 111 were from the Global South, the majority of whom were from China [260]. The entire African continent was represented by only a handful of male scientists from South Africa. Now, one can argue whether Reuters' selection criteria, which followed the criticizable but common quantitative academic citation indices, resulted in a somewhat skewed picture. De facto, however, the list broadly reflects the reality of academia. In Germany, too, only slightly more than a quarter of all professorial positions are held by women [261].

Kristina Lunz, who founded the Center for Feminist Foreign Policy in Berlin together with two women, writes a chapter on the connection between the climate crisis and feminism in her book *The Future of Foreign Policy is Feminist*: "No climate justice without feminism [262]." Undoubtedly, reasons can be found for this argument, and Lunz lists them: The voices of courageous women who have been campaigning for climate protection for years must be heard; unilateral, nationalistic approaches cannot solve the global crisis; the climate crisis is an outgrowth of an exploitative system; human security does not only mean the absence of hot conflicts; climate deniers are often also misogynists.

In fact, however, the core logic should be: no feminism without climate protection. If the foundations of life on our planet are destroyed by climate impacts, it is normative values that will be the first to go overboard. The defense of democracy, the closing of the inequality gap, and the emancipation our mothers and grandmothers fought for—these are all at risk when extreme weather conditions determine our daily lives. The extent to which issues of social justice recede into the background of public discourse during acute crises can be seen not least in the security policy shock triggered by the Russian Federation's war of aggression against Ukraine in violation of international law, which dominates the societal debates in many countries.

Part of the truth is also that at present, it is mainly men who hold executive positions in companies that are responsible for the majority of global emissions. Saudi Aramco, Gazprom, and the Iranian National Oil Company left aside—not only there, but also at Chevron, ExxonMobil, BP, and Shell, only men occupy the chairman or CEO positions. Even if there are now isolated women on some "boards," the highest top positions remain occupied by men alone. This is no mere snapshot. Rather, these are structures created by men, which—in complicity with the consumers—drive the destabilization of the world climate and therefore foster the causes of displacement. Although this is not the primary goal of these companies, it is a side effect of their way of doing business that has been known for decades and is at least tolerated—an unacceptable collateral damage.

ExxonMobil's early climate change studies in the 1980s, for example, show that it deliberately obscured for years how harmful the fossil fuel industry's products were [263]. "Instead of working on a solution, [ExxonMobil] invested millions of dollars in PR campaigns to sow doubts about climate science," writes the German climate scientist Stefan Rahmstorf, pointing to the accusations of the daughter of one of ExxonMobil's leading scientists, who publicly rebelled against the company [264, 265].

Aggressive lobbying, such as by the oil company BP, which in 2017 supported the inauguration of US President Donald Trump with half a million dollars and was rewarded for this with environmental deregulation, is now leading to extensive corporate greenwashing. This is more of a desperate last stand for their own right to exist than an honest restructuring. Such a restructuring would probably also include women taking on top positions, even if it is not certain that women would bring about public welfare-oriented emission reductions. But given the steadily declining emissions budget and the snail-like pace of improving gender equality, it is unlikely that women will move into leadership positions at energy companies in the near future and demand compliance with the Paris temperature limits. According to the World Economic Forum's "Global Gender Gap Report," it would take nearly 136 more years for men and women to achieve global equality under the current trajectory [266]. Women are already active as agents of change. The shaky foundations on which the fossil patriarchy now stands can be seen not least in the Fridays for Future movement of schoolchildren, in which a number of schoolgirls (such as Greta Thunberg, now an adult) stirred the masses and brought some coal bosses to their knees through their strong-willed protest.

The courageous resignation of a long-time security advisor to the oil company Shell, Caroline Dennett, is also part of the female resistance to the fossil fuel lobby. Dennett publicly declared that she would stop working with the oil giant because, despite its zero-risk strategy, according to her, Shell was paying no attention to climate risks and even wanted to develop new sources of fossil fuels. "I can no longer work for a company that ignores all the alarm signals and denies the risks of climate change and ecological collapse," Dennett wrote on her LinkedIn page in 2022, sending a letter to that effect to all the company's executive directors. Immediately, hundreds of men in the comment columns became outraged and berated Dennett.

Despite the hatred, courageous women around the world are standing up to the climate and biodiversity crisis. Alessandra Korap Munduruku fights against the deforestation of the Amazon rainforest and for the rights of the indigenous population in Brazil. Because she pillories large agribusinesses and soy production, she often receives death threats. "We are living in a moment of great shock," says the representative of the Munduruku people, who has won several awards for her courage.

Mary Robinson, former President of the Republic of Ireland and former UN High Commissioner for Human Rights, has long advocated for climate justice: "climate change denial is not just ignorant, it is malign, it is evil, and it amounts to an attempt to deny human rights to some of the most vulnerable people on the planet [267]." When I approached her at a conference more than a decade ago and told her about my dissertation project on climate migration, she advised me to put the affected people at the center of the work and not to write a purely theoretical treatise. I am grateful to her for this advice to this day.

Since the course for a stable global climate must be set today, it is not enough to simply wait until the younger generation moves up to positions of responsibility. While women are not inferior to men in purely numerical terms, they do not have an equal share of power, even if individual women do hold positions of power. The fact that men are disproportionately often found in positions deciding the future of all genders is unquestionably unfair and must change. To have a steering effect, climate protection and feminism must be advanced equally and in parallel. Decisions made by men in industrial companies, governments, and climate delegations have significantly shaped the course of the climate crisis. Since the climate crisis needs to be solved immediately, women as well as young people in general depend on men in positions of power to act on their behalf for the common good.

Men's solidarity in the twenty-first century thus no longer means rescuing women and children from the sinking ship first but preventing it from sinking. That is far more complex. Fortunately, many male scientists, journalists, activists, teachers, meteorologists, entrepreneurs, investors, and so forth have already declared war on the climate crisis. More voices are needed.

A Social Problem

A crucial conclusion emerges from the different dimensions of the justice issues associated with climate migration: If we perceive climate change as a purely environmental problem, we will fail to find effective solutions. It is a human rights issue, a gender crisis, and an inequality catastrophe, a foreign and domestic policy problem that threatens the security of our countries. Climate risks affect urban planning, architecture, and world heritage. They affect the drivers of migration and displacement and impact host countries, refugee camps, and informal settlements. Without this overarching understanding, putative solutions can lead to even greater marginalization of certain groups. Therefore, it is important to learn from the lived realities of those affected and to incorporate their experiences into policy reforms and concrete projects.

Over the years of my research, one realization came to me: The stories of people like Renate from the Ahr Valley, Moyna from Bangladesh, Kilok from the Marshall Islands, and Traoré from Burkina Faso are highly complex and individual, but they also have a lot in common: the caesura of displacement, defenselessness in the face of unleashed forces of nature, the will to survive, and uncertainty about the future are part of their life stories. At the same time, the people affected have very different options at their disposal for shaping their lives after the extreme weather events.

It is therefore necessary to make the call from different parts of the world for more climate justice be heard and to become active at the local, regional, and global level. Political and legal reforms, as discussed in the previous chapters, are just as necessary as concrete, immediate help. Both are increasingly demanded by activists who think globally and see their own future as linked to the future of the Marshall Islands and Bangladesh. In the meantime, a world citizen movement has emerged [268], carried by students who strike on Fridays, often supported by teachers, scientists, parents, and grandparents. They all recognize the warning signs in the Earth system and appreciate the value of science in public discourse. This is new and leaves some room for a confident outlook on the future.

Solutions Outside the Box

Far too seldom do we look at the approaches to solutions and pioneering work in countries outside Europe and the United States. An even closer exchange, for example, at the municipal level, would be desirable so as not to reinvent the wheel after every storm. An impressive case of fun educational programming is the work of the Mediae Company, which has established various successful formats in East Africa to impart knowledge. For example, there is "Shamba Shape Up," a reality TV series that visits several small farms in East Africa. "Shamba" is Swahili and means field, farmland, or farm, so "Shamba Shape Up" means something like "Pimp my farm." The farmers talk about their agricultural challenges, and experts give advice on what they can improve. They also provide practical help like for repairing stables. In addition to information on sustainable agriculture, there are tips on nutrition, hygiene, and other topics. All of this generates widespread interest, and the series is extremely successful.

Another format of the social enterprise is the soap opera Makutano Junction, which reaches millions of viewers. With entertaining drama, serious social issues such as social justice, environmental pollution, corruption, and diseases like HIV are addressed, and possible ways out of the problems are shown.

Even if this portfolio cannot be transferred 1:1 to the realities of life in other countries, it would be entirely conceivable to use similar formats to spread knowledge about disaster and infection prevention or how to cope with the climate crisis. In this way, information could be conveyed more popularly and thus perhaps more successfully than before.

Architects for the Poorest

If climate change forces millions of people to seek new homes in the coming decades, the pressure on megacities will likely increase. As noted earlier, this also affects metropolitan areas that already have high population densities with weak governance. In the Philippine capital Manila, for example, slums now make up large parts of the urban area.

In Lagos, the capital of Nigeria, a huge slum has grown into the sea over the years. Some of the dwellings stand on stilts—a colorful sight from afar, but in reality they are emergency shelters in which the residents literally struggle to keep their heads above water. A few years ago, a floating school has been built in Makoko, as the informal settlement is called, based on designs by Nigerian architect Kunlé Adeyemi. Whether in Asia, Africa, or other parts of the world, architects are challenged to develop new environmentally friendly methods and building techniques for the poorest of the poor so that they can lead a decent life. Natural materials such as earth and wood as well as renewable energies such as wind and sun play a decisive role in this.

In the past, many slums were created without urban planning, mainly by improvisation of the settlers because everything was lacking—building materials, money, and knowledge. Unfortunately, the majority of schools of architecture focus on designing innovative concepts for high-rise and high-end buildings to be inhabited by the global upper middle class. The real challenge, however, is to create appropriate and affordable housing that also fosters community and inclusion. Therefore, architects and urban planners are faced with the task of exploring new construction techniques and training the next generation on how to design and realize projects with extremely limited resources—materials, finances, and area—in different geographical and cultural spaces.

Architects Yasmeen Lari from Pakistan and Kunlé Adeyemi from Nigeria, for example, are pursuing such solutions and implementing them together with local people [269]. Expanding such collaborations between architects and local practitioners would be desirable and could promote adaptation to environmental change and disaster resilience when integrating urban agriculture projects into urban slums. Ideally, residents would receive vocational training in specific techniques during the implementation phase, such as building houses with earthbags [270]. There are already successful pilot projects for this, which need to be supported by universities, local administrations, and development agencies. Subsequently, these projects can be adapted to different local conditions to ensure systemic change in this way.

Floating Cities as a Lifeline?

Kunlé Adeyemi's work is groundbreaking for many areas because by the end of the twenty-first century, sea level experts such as Stefan Rahmstorf from Germany expect sea levels to rise by up to more than 1 m in a scenario of unchecked CO_2 emissions [271]. This would mean "land under the sea" for coastal regions, the disappearance of individual islands, and the demise of entire flat-lying island states. Currently, 110 million people live on land areas that lie below the current high tide line [272]. If sea levels rise, these areas will expand, and the pressure on infrastructure designed to hold back the water will increase.

Even if the Paris Climate Agreement is adhered to, sea levels could rise over half a meter by 2100. In addition to architects, also marine scientists, engineers, landscape planners, and builders are now looking for solutions to give people from these endangered areas a future in their own country. And "in their own country" in this case means: on their own water near the coast, river deltas, waterways, and lagoons.

Like Adeyemi's floating school, floating houses, and aqua architecture in general are gaining more and more followers. The densely populated Netherlands, a quarter of which is below sea level, and which has been fighting the tides of the North Sea since time immemorial, can also call itself a pioneer in this field. For many years, the Netherlands has been experimenting with ways of living on the water [273]. The idea is not just to build leisure and prestige properties such as houseboats, vacation homes, or water villas, but also housing complexes for thousands of average citizens. In the Netherlands, complete floating settlements have already been built in this way, such as IJburg in Amsterdam or Maasbommel near Arnhem. Some of the houses can rise several meters because they sit on a float, a hollow concrete tube which can slide up [274]. Other green technologies such as heat pumps, solar panels, wastewater treatment systems, green roofs, and rainwater collection tanks are designed to make the homes largely independent and promote sustainability.

Naturally, this type of urban design is also attracting great interest from other countries that are particularly threatened by rising seas, such as the Philippines, island states, or metropolises located on the waterfront. Another example is South Korea's second-largest city, Busan, which in April 2022, together with UN-Habitat (the United Nations program for human settlements) and the company Oceanix, presented its Oceanix Busan project in New York—a floating small city on three main platforms for 12,000 people [275]. The architectural firm of the Dane Bjarke Ingels created the innovative designs of the "Oceanix City [276]." Core to the idea is the implementation of several Sustainable Development Goals. This includes assessing energy consumption and natural resource needs over the life cycle of the buildings. Moreover, renewable energy, sustainable mobility, and circular resource concepts are envisioned. Construction is planned from 2025 to 2028.

The Imperfect City of Tomorrow

Although floating cities inspire awe, it is unlikely that such costly constructs will be widely implemented in the foreseeable future. Much of the research on sustainable urban planning and design to date has focused on imagining concepts for the ideal city, characterized by high-end technologies that serve the well-being of its inhabitants while safeguarding the services of the Earth system [277].

However, many of these innovations are not viable options for the majority of the world's population in this century, certainly not for people affected by poverty. Challenges such as climate change, population growth, and rural-urban migration require a different approach. The associated problems are particularly demanding on the administrations of large cities, and the need for action is enormous—both in terms of emissions reductions and climate adaptation.

Among other things, the "Bauhaus Earth," an alliance of science, architecture, urban planning, and forestry, wants to meet these challenges. The project, which is based on the famous model from the 1920s that strives for renewal in architecture, the fine arts, and design, also aims for revolutionary change in many respects. New and old materials are to be used and, in particular, one central element; wood from sustainable cultivation. Long-lasting wooden buildings form CO_2 sinks, and cement and concrete, which cause large amounts of greenhouse gas, are to be comprehensively replaced. Scientist Galina Churkina has shown that this is possible. Together with an international team of authors, including myself, she investigated how much CO_2 can be stored by timber construction and how this compares with the greenhouse gas balance of conventional building methods [278]. The results were astonishing: On the one hand, a massive amount of CO_2 is saved and stored, and on the other hand, there is enough wood to build much more from the raw material. If 90% of houses were built from wood by mid-century, up to 20 gigatons of carbon could be sunk into buildings.

The "Bauhaus Earth" is now growing from an idea to an institution, driven by climate scientist Hans Joachim Schellnhuber and architect Philipp Misselwitz. The plan to *build* our way out of the climate crisis is also finding favor at the European level. European Commission President Ursula von der Leyen, for example, proclaimed the New European Bauhaus. She formulated the project's ambition as follows: "We want to create livable and affordable housing for more and more people—and at the same time protect the climate and the environment [279]."

Climate Knowledge and Culture

The consequences of climate change are also being addressed more and more in the cultural sector. For example, the German theater maker Michael Ruf and his nongovernmental organization "Wort und Herzschlag" (Word and Heartbeat) bring the life stories and suffering of various people to the stage as monologues that branch out into one another. What is special about this is that Ruf creates the scripts from many hours of interview material, which he merely condenses, but otherwise retains verbatim. This results in immediately moving documentary texts. His work touches and shakes people. After the Mediterranean Monologues, which dealt with the fate of refugees and the efforts of courageous helpers, he turns to the consequences of climate change in his latest piece, for which he traveled to Bangladesh to talk to displaced people there. At a meeting in Berlin, Michael Ruf explained to me: "Although there is a lot of talk about climate change in this country, very few people know what it actually means for people in Bangladesh, Mozambique or Tanzania. With the Climate Monologues, I want to show how climate change is already turning the lives of many people upside down in other regions of the world."

Another project based in Germany combines work with refugees with efforts to improve climate protection. The initiative "KlimaGesichter" (Climate Faces), funded by the German Federal Ministry for the Environment, trains refugees to become climate ambassadors. The goal is to provide them with knowledge about climate change and climate protection that they can apply in their future jobs, whether in Germany or after returning home. The project aims to combine refugee's integration into society and climate protection.

Spring Heat Waves

In April 2022, I am once again in New Delhi, where my journey on climate migration started. Although it is spring, the country has already experienced several heat waves, beginning as early as March. When I arrive, temperatures climb to levels between 40 and 45 °C for several days. It is unbearably hot outside. Later, the German climatologist Friederike Otto and her attribution research team will find that this heat wave, which will also affect Pakistan, has been made 30 times more likely by previous global warming [280]. In contrast, it is pleasantly cool in the conference rooms of the Raisina Dialog, which I am attending, a geopolitical meeting led by the Indian Ministry of Foreign Affairs and the Observer Research Foundation, an Indian think tank.

Up for debate in the session I am attending are issues of CO_2 pricing—Who has to pay for the emissions?—and the lack of help for those who do not even have an electricity connection. Our round of talks goes harmoniously despite partly different views—this kind of exchange is irreplaceable in order to be able to gain external views on Western politics. And the issues we discuss could hardly be more topical because outside the asphalt is practically burning. It is obvious: Climate change increases inequalities. While the middle and upper classes are cooling themselves down 24 h a day with air-conditioning in order to be able to sleep and work, the poorest sections of the population are at the mercy of the heat. In some cities, air-conditioned halls have been set up so that workers can at least get a few hours of sleep, preventing the number of heat-related deaths from skyrocketing.

Meanwhile, in the plenary hall, a high-level panel is taking place. Former Environment Minister of the Maldives, Aminath Shauna; then Foreign Minister of Norway, Anniken Huitfeldt; the Secretary-General of the OECD, Mathias Cormann; and Amitabh Kant, then CEO of the Indian planning and consulting agency NITI Aayog, talk about the global energy transition [281]. As the discussion begins, it becomes clear how far apart from one another the speeches are. To the Indian moderator, who wants to know how Norway could support developing countries in the energy transition, then Norwegian Foreign Minister Huitfeldt replies, "We are an oil-producing nation and also producing a lot of gas and we will develop our petroleum sector." She further on explains that they will, however, also invest in renewable energies with profits from the oil and gas revenue, but "not support a dismantling of the Norwegian gas and oil producing sector."

Shauna, then Environment Minister of the Maldives, on the other hand, outlines the extreme challenges facing the flat-lying island nation: "The Maldives is 1200 islands and the highest point in our islands is just about 1 m. [...] So, any increase in sea level is an existential threat for us. Our islands are eroding at a much faster rate than we had imagined. Our islands are getting flooded every year. We are losing our country to climate change." When asked about the biggest challenge in the global energy transition, she replied, "I don't think it's the lack of finance. I also don't think it's the lack of technology. I think it is the lack of political will [...]. We are not treating the climate crisis like an emergency."

Aminath Shauna's statement seems to fall flat in the hall, with no one really addressing what she says. Instead, the OECD Secretary-General later emphasizes in his statement that the transition to renewable energies will take several decades. The OECD countries in particular, however, would have to exit the fossil fuel industries more quickly than developing countries. This has been discussed for a good 30 years now, while the inhabitants of some island states are literally up to their necks in water. But the statement is hardly surprising, since Cormann, in his previous role as Australia's finance minister, already called net-zero emission reduction targets "extremist" and repealed CO_2 pricing in Australia in favor of the coal lobby [282].

When the Q&A session for the audience opens, I turn to Cormann. I am interested in what is being done this year to reduce emissions, since climate impacts like the heat wave that is surrounding us in India right now will not allow another delay by decades. What follows is my personal "Don't Look Up" moment. The OECD Secretary-General, in apparently practiced manner, ridicules my statement and claims that I probably want to shut down all power plants overnight. This tactic of delegitimization is many decades old and yet somehow shocking in this setting. It aims to exaggerate obviously necessary measures, such as taking the first steps in reducing emissions, so that they seem completely absurd.

Disillusioned and full of doubt, I sit down again. Then, suddenly an older man approaches me from the side, dressed simply compared to the conference glamour. "Do you live here in Delhi?" he asks. "No," I reply in a whisper, as the panel continues in front of us, "but I used to live here for a few months." "I see," he says, pausing. "Because you speak from the heart of us, the ordinary people here." Shortly after, he was gone again.

I have often thought of this exchange since then. The panel discussion was exemplary of how compassion for the other person is lost in many debates. My short conversation with the stranger, on the other hand, was the opposite. The trip to India made me realize once again that we are living in a time of multiple crises. After 2 years of the Covid-19 health emergency, many hoped to breathe a sigh of relief. But then the incomprehensible happened: Russia launched a war of aggression against Ukraine on February 24, 2022. The discussions at the Raisina Dialogue were also marked by the cracks in the international order. The war in the middle of Europe is driving hundreds of thousands to flee; thousands of people are dying, among them many children. The crisis also hit the European Union at its Achilles' heel in terms of energy policy and its dependence on coal, oil, gas, and uranium from Russia. Will lessons be learned from the geopolitical and humanitarian catastrophe for a security policy that understands resource and energy issues as part of internal security?

What raises hope is a look at the solar and wind atlas that the World Bank is compiling with research institutions such as the Technical University of Denmark (DTU). It shows the enormous potential of renewable energies [283]. As Shauna, the then Environment Minister from the Maldives, pointed out, there is no shortage of options.

In the long run, given the abundance of possibilities, it will be difficult to justify people losing their homes, their properties, even loved ones, because the society at large is unwilling to shift energy systems more quickly or consume less meat. If migration is proclaimed as a form of adaptation to climate change, we cannot close our doors to those who must leave their homes behind because of the lifestyle changes we refuse to make. The fossil fuel system will not remain stable in the long run. But how do we make the change? And how do we do it inclusively and limit painful disruptions?

The migration of many people I spoke with began with the hope that there would be less adverse conditions elsewhere than in their place of origin. They thought about what advantages and disadvantages migration might bring. While many could not fully assess what the change of location would entail at the time of migration, the pressures of climate change and other life circumstances were often such that they chose to face this very uncertainty. In some cases, however, this was not a choice for a better life, but a choice between life and death.

This balancing process holds important lessons for overcoming the climate crisis. While the first steps we need to take are obvious, the path to solving the crisis is not mapped out. It will be marked by contradictions in our actions and will also involve failures. But fearing the tenth step, we must not lose sight of the need to take the first, second, and third steps. Even against the backdrop of uncertainty, we must make decisions that at least steer us in the right direction. One thing is certain: The fight against the climate crisis ends either with its solution or with our capitulation.

What Is Next? Judgment, Conviction, and Empathy

Will we succeed in stabilizing the global temperature at a maximum of 2 °C and thus protecting essential habitats? The answer must remain open. Time is running out, but the faster we cut emissions now, the greater the chances that humanity will find a way out of this crisis as well. If we prolong our hesitation, more and more decisions will be taken out of our hands.

Migratory movements are the sum of many individual stories. To prevent these movements of people from becoming a humanitarian crisis, we need diverse approaches, new ideas, and more people to address the complex connections between climate impacts and their effects on livelihoods and migration. Away from ideology-driven debates, real change can grow. There will be no single solution, but the bouquet of options still open to us is lush. If we understand migration as an act of liberation from the oppressive force of the climate crisis, there is no way around assuring greater freedom of movement and enshrining more rights as people flee the forces of nature. For migration as the source of collective hope is part of our human history.

Now I would like to invite you to conclude by taking another look at the theses summarized at the end of each chapter:

1. The climate crisis has become the new normal, even if its extent can still be limited. Migration enables those affected to survive, but not necessarily to maintain their standard of living.
2. The existing border regime will lead to increasing human rights violations in the wake of dramatic climate impacts. Unevenly distributed environmental harm requires institutional and legal reforms that open safe, legal migration routes from degraded areas and limit future damage.
3. On the small island states, the struggle for survival in the climate chaos has already begun. The imminent demise of entire cultures is being accepted for the growth paradigm of the fossil industries. The islands without a future are sending out a warning signal—a last call to avert the global catastrophe.
4. The pressure at Europe's external borders will continue to increase, and European values will have to be measured by how much human suffering the EU is willing to accept in favor of border protection.
5. Successive superstorms could depopulate exposed areas in the long term. Progress in disaster management in many Asian countries is a hopeful sign of higher resilience, but advancements are competing with increasingly violent climate impacts.

6. The interplay of species extinction and climate change is pushing humanity and its habitat toward an unprecedented crisis. Only migration will allow certain species to survive.
7. Even in Central Europe, people may lose their homes and cultural assets due to climate impacts. Although in global comparison, more options for adaptation are available, rising emissions mean that these opportunities are shrinking and becoming more costly. The cheapest solution remains comprehensive and effective climate protection.
8. A wealth of tools and recommendations to protect climate displaced people is waiting to be implemented. Drawing on existing knowledge and mustering the political courage for new directions could improve the prospects for many people whose future will be determined by the climate crisis.

The thesis from this chapter is as follows:

9. Even if the same opportunities are not open to all people, at least those who significantly contribute to greenhouse gas emissions have the opportunity and moral duty to contribute to the solution of the climate crisis and to support migration from destroyed areas, through behavioral changes, participation in elections, protest or resistance.

In addition to the many proposals already mentioned on how to overcome the human crisis of climate change, there are three things in particular that will get us there: First is *discernment* of what is right and what is wrong and especially judging what is essential and what is dispensable. Second is the *conviction* that we will muster the strength to extricate ourselves from this crisis. And third is *empathy* for people whose lives have been restricted by climate change. Standing up for their rights must be part of the solution. These are the challenges of the climate crisis that science cannot meet for us—we must do it ourselves.

Although some processes have already slipped away from us, species have become extinct, ecosystems have been destroyed, and people have been displaced, the last remaining threads of the climate web are still in our hands. On them hangs nothing less than the fate of humanity. The fact that we still have this chance and know about it—more than ever before—is a glimmer of light in the darkness of these times.

Acknowledgments

This book is based on many years of research in which various people have crossed, accompanied, and prepared my path. I would like to take this opportunity to express my special thanks to them.

People have given me their time and shared their stories in well over a hundred interviews as part of my academic work. Many special experiences and encounters are linked to this, and I am deeply grateful for this gift of life.

Franziska Günther from the Graf & Graf agency got the ball rolling for this book and gave me important advice in all phases of its creation. I would like to express my sincere thanks for this attentive support and trusting collaboration.

I would also like to thank Stefan Ulrich Meyer from dtv, who was instrumental in supporting this project and gave me important structural advice from his wealth of experience, especially at the beginning, which is now reflected in this book.

I would like to thank Rosemarie Mailänder for her excellent, patient editing and encouraging words. I could not have done better for this proofreading. She encouraged me to read the text from a different perspective.

Many people have read and commented on parts of this book at various stages. I would like to thank Dana Schirwon (Chap. 2), Mechthild Becker (Chaps. 3 and 5), Shana Tabak (Chap. 3), Stefanie Wesch (Chap. 4), Michael Ruf (Chap. 5), Jonas Bergmann (Chap. 6), Valerie Köcke (Chap. 6), Josef Ludescher (Chap. 6), Renate Petry (Chap. 7), Larissa Rausch (Chap. 7), and Annika Mannah (Chap. 7), Sima Bulut (Chap. 8), Anna Sperber (Chap. 9).

I would also like to kindly thank:

Manolo Ty for serious, beautiful, and multifaceted photographs and many helpful comments on the content.

Arne Dunker for the joint, intensive trip to Switzerland as well as comments and advice on Chap. 7.

Sven Plöger for the encouragement to tackle this project and many conversations about the climate crisis and its solution.

Prof. Michael Daxner, who has always been an attentive contact and advisor for me.

Stephan Zöllner from Diakonie Katastrophenhilfe for his trust and commitment.

Angelika Nikionok-Ehrlich, Peter Ehrlich, and David Wortmann for a final review.

My special thanks go to Prof. Hans Joachim Schellnhuber, who has accompanied my work since 2014 and shaped many of the ideas in this book. Without his pioneering scientific and humanistic work and his personal commitment, many climate policy decisions would not have been realized.

The contents of this book are based on several years of scientific work at the Potsdam Institute for Climate Impact Research, which was supported by the institute and research department management as well as various external sponsors. My sincere thanks go to them.

I am grateful to Julia Blocher, Himani Upadhyay, Dr. Emanuela Paoletti, Dr. Roman Hoffmann, Dr. Maria Martin, and Dr. Barbora Sedova for intensive discussions on climate migration.

I would like to thank Prof. Helga Weisz for supervising my doctorate, on which this book is based, and for giving me the space for scientific development.

Since I took over the management of the Center for Climate and Foreign Policy at the German Council on Foreign Relations in 2021, I have been supported by an excellent team. Thank you Tim Bosch, Dana Schirwon, Kai Kornhuber, and Leonie Oechtering for your encouragement and passion in the fight for solutions to the climate crisis.

From my research stays, I would like to thank Dieter Paulmann, founder of the Okeanos Foundation, who paved the way for me to the Marshall Islands. On site, Dustin Langidrik, Iva Nancy Vunikura, Alex Tohitika Sanchez, Alexander Beetz, and Raffael Held helped me a lot.

I would like to thank Dr. Dorothea Rischewski and Teddy Fong for our joint work on climate migration on island states and Claire Frank for her support on Barbuda.

I thank Prof. Rainer Sauerborn for an unforgettable research trip to Burkina Faso.

I would like to thank Fabian Wolff, Naoshin Jahan, and Rashed Jalal Himal for supporting my work in Bangladesh.

Josef Schuler and the Infanger family showed us the beautiful Swiss Isenthal, thank you for your hospitality.

For help with the English version of this book, I would like to deeply thank Prof. Thomas Pogge, Lindsay Visser, and Dr. Viney Setya. A special thanks goes to Aaron Schiller and Ragavendar Mohan from Springer and Andrea Seibert from dtv for their support and patience.

I would like to thank my parents, Barbara Lange-Vinke and Hermann Vinke, to whom this book is dedicated and who have always stood by my side, including during this project.

"Without love, all is nothing." Thank you, Thiago.

Bibliography

1. Black, R. / Bennett, S. R. G. / Thomas, S. M. / Beddington, J. R.: Migration as Adaptation. Nature 478 (2011).
2. Climate neutrality refers to a "Concept of a state in which human activities result in no net effect on the climate system" (IPCC, 2018: Annex I: Glossary [Matthews, J.B.R. (ed.)]. In: Global Warming of 1.5°C. An IPCC Special Report on the impacts of global warming of 1.5°C above pre-industrial levels and related global greenhouse gas emission pathways, in the context of strengthening the global response to the threat of climate change, sustainable development, and efforts to eradicate poverty [Masson-Delmotte, V., P. Zhai, H.-O. Pörtner, D. Roberts, J. Skea, P.R. Shukla, A. Pirani, W. Moufouma-Okia, C. Péan, R. Pidcock, S. Connors, J.B.R. Matthews, Y. Chen, X. Zhou, M.I. Gomis, E. Lonnoy, T. Maycock, M. Tignor, and T. Waterfield (eds.)]. Cambridge University Press, Cambridge, UK and New York, NY, USA, pp. 541-562, https://doi.org/10.1017/9781009157940.008).
3. Vinke, K.: Unsettling Settlements. Cities, Migrants, Climate Change. Lit Verlag (2019).
4. Lenton, T. M. et al.: Climate tipping points— too risky to bet against. Nature 575 (2019).
5. IPCC, 2021: Climate Change 2021. The Physical Science Basis. Contribution of Working Group I to the Sixth Assessment Report of the Intergovernmental Panel on Climate Change. [Masson-Delmotte, V. / Zhai, P. / Pirani, A. / Connors, S. L. / Péan, C. / Berger, S. / Caud, N. / Chen, Y. / Goldfarb, L. / Gomis, M. I. / Huang, M. / Leitzell, K. / Lonnoy, E. / Matthews, J. B. R. / Maycock, T. K. / Waterfield, T. / Yelekçi, O. / Yu, R. / Zhou, B. (Hrsg.)]. Cambridge University Press, in Druck, 2021.
6. UNHCR: Global Trends – Forced Displacement in 2020 (2021).
7. Hoffmann, R. / Dimitrova, A. / Muttarak, R. / Crespo Cuaresma, J. / Peisker, J.: A meta-analysis of country-level studies on environmental change and migration. Nature Climate Change (2020); https://doi.org/10.1038/s41558-020-0898-6.
8. Schellnhuber, H. J.: Selbstverbrennung – Die fatale Dreiecksbeziehung zwischen Klima, Mensch und Kohlenstoff. C. Bertelsmann (2015).
9. IDMC (2024): GRID 2024 – Global Report on Internal Displacement. Internal Displacement Monitoring Centre and Norwegian Refugee Council.
10. The Report analyzed the following regions: SubSaharan-Africa (86 Millionen climate related internal displacements), North-Africa (19 million), East Asia and the Pacific (49 Millionen), South Asia (40 million), Latin America (17 Millionen), Central Asia and Eastern Europe (5 million). Source: Clement, V. / Rigaud, K. K. / de Sherbinin, A. / Jones, B. / Adamo, S. / Schewe, J. / Sadiq, N. / Shabahat, E. (2021). Groundswell Part 2: Acting on Internal Climate

Migration. World Bank, Washington, D. C., World Bank. https://openknowledge.worldbank.org/handle/10986/36248; License: CC BY 3.0 IGO.
11. Bengtsson, L. / Lu, X. / Thorson, A. / Garfield, R. / von Schreeb, J.: Improved response to disasters and outbreaks by tracking population movements with mobile phone network data: a post-earthquake geospatial study in Haiti (2011). PLOS Medicine, 8(8):e1001083; https://doi.org/10.1371/journal.pmed.1001083;
12. Lai, S./Erbach-Schoenberg, E.z./ Pezzulo, C. et al. Exploring the use of mobile phone data for national migration statistics. Palgrave Commun 5, 34 (2019). https://doi.org/10.1057/s41599-019-0242-9.
13. Danuor, S. et al.: Education in meteorology and climate sciences in West Africa. Atmospheric Science Letters. Wiley Online Library (2011).
14. World Bank: CO_2 emissions (metric tons per capita) – Sub-Saharan Africa (2018); https://data.worldbank.org/indicator/EN.ATM.CO2E.PC?locations=ZG.
15. Vinke, K. / Harper, A.: Climate Change and the future of sage returns (2020); https://reliefweb.int/sites/reliefweb.int/files/resources/Climate%20change%20and%20the%20future%20of%20safe%20returns%20%28November%202020%29.pdf.
16. Khanna, P.: Move - How Mass Migration Will Reshape the World - and What It Means for You, W&N, (2022).
17. Tiihonen, J. / Halonen, P. / Tiihonen, L. et al.: The Association of Ambient Temperature and Violent Crime. Scientific Reports 7, 6543 (2017); https://doi.org/10.1038/s41598-017-06720-z.
18. Upadhyay, H. / Vinke, K. / Weisz, H.: »We are still here« – Climate change and immobility in highly mobile Himalayan communities. Climate and Development, 16(5), 443–457. https://doi.org/10.1080/17565529.2023.2230176.
19. Knaus, G.: Welche Grenzen brauchen wir? Piper (2020), S. 53.
20. UNHCR: Convention and Protocol Relating to the Status of Refugees, Article 1 (1951); https://www.unhcr.org/media/convention-and-protocol-relating-status-refugees
21. Vgl. Sydney Declaration of Principles on the Protection of Persons Displaced in the Context of Sea Level Rise. https://disasterdisplacement.org/portfolio-item/sydney-declaration.
22. UNHCR: Convention and Protocol Relating to the Status of Refugees, Article 33 (1951). https://www.unhcr.org/dach/wp-content/uploads/sites/27/2017/03/GFK_Pocket_2015_RZ_final_ansicht.pdf.
23. Rankin, J.: Head of EU border agency Frontex resigns amid criticisms (2022). https://www.theguardian.com/world/2022/apr/29/head-of-eu-border-agency-frontex-resigns-amid-criticisms-fabrice-leggeri
24. Graham-Harrison, E.: »They treated her like a dog« – tragedy of the six-year-old killed at Croatian border. The Guardian (2017).
25. McAdam, J.: Protecting People Displaced by the Impacts of Climate Change. The UN Human Rights Committee and the Principle of Non-refoulement. American Journal of International Law, 114(4), 708-725 (2020); https://doi.org/10.1017/ajil.2020.31.
26. Ibid.
27. Behrouz Boochani describes in his book <No friend but the mountains> the brutal reality of an internment camp in Papua-New Guinea. As a Kurdish Iranian refugee, he spent six years in the camp on Manus Island and wrote his book in secret via cell phone text messages.
28. United Nations: Global Compact for Safe, Orderly and Regular Migration (A/RES/73/195) (2018).
29. United Nations: Global Compact on Refugees. A/73/12, Part II (2018).
30. Nansen Initiative: Agenda for the Protection of Cross-Border Displaced Persons in the Context of Disasters and Climate Change (2015).
31. For a more detailed overview of people's rights throughout the migration cycle see: IOM Outlook on Migration, Environment and Climate change (2014); https://publications.iom.int/system/files/pdf/mecc_outlook.pdf.
32. Ionesco, D. / Mokhnacheva, D. / Gemenne, F: Atlas der Umweltmigration. Oekom Verlag (2018).

33. Thornton, F.: Climate Change and People on the Move. International Law and Justice. Oxford (2018).
34. UN Human Rights Council, 48th session (2021); https://undocs.org/a/hrc/48/l.23/rev.1.
35. Bundesverfassungsgericht, Press Release (2021); https://www.bundesverfassungsgericht.de/SharedDocs/Pressemitteilungen/DE/2021/bvg21-031.html.
36. Bundesverfassungsgericht, Beschluss vom 24.3.2021; https://www.bundesverfassungsgericht.de/SharedDocs/Entscheidungen/DE/2021/03/rs20210324_1bvr265618.html.
37. Verheyen, R. /Endres, A.: Wir alle haben ein Recht auf Zukunft: Eine Ermutigung. dtv (2023).
38. For the latest developments regarding this case see: https://rwe.climatecase.org/.
39. People's Climate Case (also called Carvalho Case) C565/19 P, https://curia.europa.eu/juris/document/document.jsf?text=&docid=239294&pageIndex=0&doclang=EN&mode=req&dir=&occ=first&part=1&cid=504038.
40. People's Climate Case: EU Court turn a deaf ear to citizens hit by the climate crisis (2021); https://peoplesclimatecase.caneurope.org/2021/03/eu-court-turn-a-deaf-ear-to-citizens-hit-by-the-climate-crisis/.
41. https://www.noaa.gov/explainers/deepwater-horizon-oil-spill-settlements-where-money-went.
42. Kujawinski, E. B. / Reddy, C. M. / Rodgers, R. P. et al.: The first decade of scientific insights from the Deepwater Horizon oil release. Nat Rev Earth Environ 1, 237–250 (2020); https://doi.org/10.1038/s43017-020-0046-x.
43. Climate Accountability Institute: Carbon Majors (2020). https://climateaccountability.org/carbonmajors_dataset2020.html.
44. Global Witness: A Bad Year for Glencore (2018); https://www.globalwitness.org/en/blog/bad-year-glencore/.
45. 20th Congress of the International Criminal Law Association, 15.11.2019.
46. International Law Commission: Summary Records of the meetings of the 38th session, 119–121, 128, 175, U. N. Doc. A/CN.4/SER.A/ (1986); https://legal.un.org/ilc/publications/yearbooks/english/ilc_1986_v1.pdf.
47. Legal theorist Georg Jellinek (1851-1911) defined the three elements of a state as a "state's territory, state's people and state's power" in his standard work 'Allgemeine Staatslehre'. The Montevideo Convention on the Rights and Duties of States (1933) also defined a fourth element, the ability to conduct diplomatic relations with other countries. This principle has not been established in practice.
48. Storlazzi, C. D. / Gingerich, S. B. / van Dongeren, A. / Cheriton, O. M. / Swarzenski, P. W. / Quataert, E. / Voss, C. I. / Field, D. W. / Annamalai, H. / Piniak, G. H. / McCall, R.: Most atolls will be uninhabitable by the mid-21st century because of sea-level rise exacerbating wave-driven flooding. Sci. Adv. 4, eaap9741 (2018).
49. McAdam, J.: »Disappearing States«, Statelessness and the Boundaries of International Law (2010). UNSW Law Research Paper No. 2010-2, verfügbar auf SSRN: https://ssrn.com/abstract=1539766.
50. Ibid.
51. Grote Stoutenburg, J.: When Do States Disappear? In: Threatened Island Nations – Legal Implications of Rising Seas and a Changing Climate. Cambridge University Press (2013).
52. Ibid.
53. McAdam, J., »Disappearing States«, a. a. O.
54. Universal Declaration of Human Rights. UN-Resolution 217 A (III) 10.12.1948.
55. Burkett, M.: The Nation Ex-Situ. In: Threatened Island Nations – Legal Implications of Rising Seas and a Changing Climate (2013).
56. The names of the interview partners have been changed for privacy reasons.
57. Okeanos Foundation: https://okeanos-foundation.org/vaka-motu/.
58. For those interested in the state of the oceans, I recommend the books by oceanographer Sylvia Earle. For example, 'Sea Change, A Message from the Oceans' (1996), which

addresses the industrialization of fisheries, and 'Oceans: A Global Odyssey' (2021), which depicts both the beauty of the oceans and the impact of humans on them.
59. Oxfam Briefing: Carbon inequality in 2030 – Per capita consumption emissions and the 1.5°C goal (2021); https://oxfamilibrary.openrepository.com/bitstream/handle/10546/621305/bn-carbon-inequality-2030-051121-en.pdf.
60. Weisgall, Jonathan M. (1994). Operation Crossroads – The Atomic Tests at Bikini Atoll. Naval Institute Press. ISBN: 1557509190.
61. Keown, M. (2017). Children of Israel: US Military Imperialism and Marshallese Migration in the Poetry of Kathy Jetnil-Kijiner. Interventions, 19(7), 930–947. https://doi.org/10.1080/1369801X.2017.1403944, https://www.si.edu/object/nmah_1303438.
62. Gerrard, M.: A Pacific Isle, Radioactive and Forgotten (Opinion). The New York Times (nytimes.com) (2016).
63. Storlazzi, C. D. et al.: Most atolls will be uninhabitable by the mid-21st century, a. a. O.
64. Price, G.: Mining India's troubled history of coal and politics. Chatham House (2021); https://www.chathamhouse.org/2021/11/mining-indias-troubled-history-coal-and-politics.
65. A group of states, primarily from developing countries, that see themselves as particularly threatened by climate change.
66. The Alliance of Small Island States, AOSIS, has 39 member states from the Caribbean, Pacific, Indian Ocean, South China Sea and Africa. All of them are small or flat-lying island states that, despite their cultural differences, have joined forces due to similar challenges imposed by the climate crisis and are committed to sustainable development.
67. https://www.bmuv.de/fileadmin/Daten_BMU/Download_PDF/Klimaschutz/paris_abkommen_bf.pdf.
68. IPCC, 2018: Global Warming of 1.5°C. An IPCC Special Report on the impacts of global warming of 1.5°C above pre-industrial levels and related global greenhouse gas emission pathways, in the context of strengthening the global response to the threat of climate change, sustainable development, and efforts to eradicate poverty [Masson-Delmotte, V. / Zhai, P. / Pörtner, H.-O. / Roberts, D. / Skea, J. / Shukla, P. R. / Pirani, A. / Moufouma-Okia, W. / Péan, C. / Pidcock, R. / Connors, S. / Matthews, J. B. R. / Chen, Y. / Zhou, X. / Gomis, M. I. / Lonnoy, E. / Maycock, T. / Tignor, M. / Waterfield, T. (Hrsg.)]. Cambridge University Press, in Druck.
69. Glasgow Climate Pact, UNFCCC: COP26 cover decision (unfccc.int) (2021).
70. Ministry of Economy of Fiji: Planned Relocation Guidelines. A framework to undertake climate change related relocation (Planned-Relocation-Guideline-Fiji-2018.pdf; adaptation-community.net), (2018).
71. Vinke, K. / Blocher, J. / Becker, M. / Ebay, J. S. / Fong, T. / Kambon, A.: Home lands. Island and archipelagic states' policymaking for human mobility in the context of climate change (2020).
72. Pörtner, H.-O. et al. (Hrsg.): IPCC Special Report on the Ocean and Cryosphere in a Changing Climate (IPCC, 2019), https://report.ipcc.ch/srocc/pdf/SROCC_FinalDraft_FullReport.pdf.
73. Our research is summarized in this report to which I refer to in this chapter: Vinke, K. / Blocher, J. / Becker, M. / Ebay, J. S. / Fong, T. / Kambon, A.: Home lands. Island and archipelagic states' policymaking for human mobility in the context of climate change (2020).
74. Micronesians in Hawaii: Migrant Group Faces Barriers to Equal Opportunity. A Report of the Hawaii Advisory Committee to the U. S. Commission on Civil Rights (2019). Hawaii Micronesian Report 2020 (usccr.gov).
75. Hawai'i State Department of Education: Resources for Families In Unstable Housing (2024). https://www.hawaiipublicschools.org/ParentsAndStudents/EnrollingInSchool/Pages/Resources-for-homeless-families.aspx. See also: United States Interagency Homelessness Council: Hawaii Homelessness Statistics (2020); https://www.usich.gov/homelessness-statistics/hi/.
76. Letman, J.; Wong, J. (2017): Hawaiians call Mark Zuckerberg 'the face of neocolonialism' over land lawsuits. TheGuardian. https://www.theguardian.com/technology/2017/jan/23/mark-zuckerberg-hawaii-land-lawsuits-kauai-estate.
77. Harlow, P.: Robert De Niro – Rebuilding Barbuda (2017, Boss Files).

78. The term "White Savior Complex" was introduced by Nigerian-American writer Teju Cole. Cole, Teju: The White-Savior Industrial Complex, 2012. https://www.theatlantic.com/international/archive/2012/03/the-white-savior-industrial-complex/254843/
79. Rodríguez-Martínez, R. E. / Medina-Valmaseda, A. E. / Blanchon, P. / Monroy-Velázquez, L. V. / Almazán-Becerril, A. / Delgado-Pech, B. / Vásquez-Yeomans, L. / Francisco, V. / García-Rivas, M. C.: Faunal mortality associated with massive beaching and decomposition of pelagic Sargassum, Marine Pollution Bulletin, Band 146, S. 201–205 (2019), https://doi.org/10.1016/j.marpolbul.2019.06.015; https://www.sciencedirect.com/science/article/pii/S0025326X19304606.
80. Wang, M. / Hu, C. / Barnes, B. B. / Mitchum, G. / Lapointe, B. / Montoya, J. P.: The great Atlantic Sargassum belt, Science 365, S. 83–87 (2019), 10.1126/science.aaw7912.
81. Klein, N.: The Shock Doctrine: The Rise of Disaster Capitalism. Penguin (2008).
82. Parts of Senegal, Mauretania, Mali, Burkina Faso, Algeria, Niger, Nigeria, Cameroon, Central African Republic, Chad, Sudan, South-Sudan, Eritrea and Ethiopia are located in the Sahel.
83. Idemudia, E., Boehnke, K.: Travelling Routes to Europe. In: Psychosocial Experiences of African Migrants in Six European Countries. Social Indicators Research Series, vol 81. Springer, Cham. (2020); https://doi.org/10.1007/978-3-030-48347-0_3
84. IOM: Missing Migrants Project, https://missingmigrants.iom.int/.
85. EU announces Sahel aid | International Partnerships (europa.eu). https://international-partnerships.ec.europa.eu/news-and-events/news/eu-announces-eu194-million-additional-support-sahel-2020-04-28_en
86. Welzer, H.: Climate Wars - What People Will Be Killed For in the 21st Century ISBN: 978-0-7456-5145-3, John Wiley & Sons, (2012)
87. C. Schleussner, J.F. Donges, R.V. Donner, H.J. Schellnhuber, Armed-conflict risks enhanced by climate-related disasters in ethnically fractionalized countries, Proc. Natl. Acad. Sci. U.S.A. 113 (33) 9216-9221, https://doi.org/10.1073/pnas.1601611113 (2016).
88. von Uexkull, N., Croicu, M., Fjelde, H., Buhaug, H.: Civil conflict sensitivity to growing-season drought, Proc. Natl. Acad. Sci. U.S.A. 113 (44) 12391-12396, https://doi.org/10.1073/pnas.1607542113 (2016).
89. Mach, K. J. / Kraan, C. M. / Adger, W. N. et al.: Climate as a risk factor for armed conflict. Nature 571, 193–197 (2019); https://doi.org/10.1038/s41586-019-1300-6.
90. C.P. Kelley, S. Mohtadi, M.A. Cane, R. Seager, Y. Kushnir, Climate change in the Fertile Crescent and implications of the recent Syrian drought, Proc. Natl. Acad. Sci. U.S.A.
91. Wesch, S./ Rheinbay, J. / von Soest, C. / Gornott, C. / Scheffran, J. / Vinke, K. (under review): Fulani Identity in Crisis.
92. Vinke, K. / Rottmann, S. / Gornott, C. / et al. Is migration an effective adaptation to climate-related agricultural distress in sub-Saharan Africa?. Popul Environ 43, 319–345 (2022). https://doi.org/10.1007/s11111-021-00393-7
93. The names of the interview partners have been changed for privacy reasons.
94. Asefa's story (name changed) and other aspects of the climate impact and rural development nexus are detailed in World Vision's publication "Ethiopia - Why Forests Mean Future." (2019).
95. Tomalka, J. / Lange, S. / Röhrig, F. / Gornott, C.: Climate Risk Profile for Ethiopia (2020).
96. Vinke, K. / Becher, M. / Fahimi, A. / Flamm, P. / Ilse, S. / Kroll, S. / Kruckow, C. / Ritzer, T. / Schalatek, L. / Scheffran, J. / Šedová, B. / Sperber, A. / Strumpf, N. / Wesch, S.: Civilian Crisis Prevention through Environmental Peacebuilding. Environmental and climate-focused approaches for sustainable peace. Published by the Advisory Board on Civilian Crisis Prevention and Peacebuilding to the German Federal Government. Study Number 7. Berlin. (2024)
97. Rinaudo, T. / McKenzie, S. / Huynh, TB. / Sterrett, C.L.: Farmer Managed Natural Regeneration: Community Driven, Low Cost and Scalable Reforestation Approach for Climate Change Mitigation and Adaptation. In: Leal Filho, W., Luetz, J., Ayal, D. (eds)

Handbook of Climate Change Management. (2021) Springer, Cham. https://doi.org/10.1007/978-3-030-22759-3_281-1
98. World Vision Australia: FMNR Manual, (2018).
99. Binam, J.N. / Place, F. / Kalinganire, A. et al.: Effects of farmer managed natural regeneration on livelihoods in semi-arid West Africa. Environ Econ Policy Stud 17, 543–575 (2015). https://doi.org/10.1007/s10018-015-0107-4
100. Pemunta, N. V. / Ngo, N. V. / Fani Djomo, C. R. / Mutola, S. / Seember, J. A./ Mbong, G. A. / et al.: The Grand Ethiopian Renaissance Dam, Egyptian National Security, and human and food security in the Nile River Basin. Cogent Social Sciences, 7(1). (2021) https://doi.org/10.1080/23311886.2021.1875598.
101. Sterl, S. / Fadly, D. / Liersch, S. et al.: Linking solar and wind power in eastern Africa with operation of the Grand Ethiopian Renaissance Dam. Nat Energy 6, 407–418 (2021); https://doi.org/10.1038/s41560-021-00799-5.
102. NOAA: Active 2021 Atlantic hurricane season officially ends (2021); https://www.noaa.gov/news-release/active-2021-atlantic-hurricane-season-officially-ends.
103. NOAA: How do Hurricanes form? National Ocean Service website; https://oceanservice.noaa.gov/facts/how-hurricanes-form.html.
104. IPCC, 2021: Summary for Policymakers. In: Climate Change 2021: The Physical Science Basis. Contribution of Working Group I to the Sixth Assessment Report of the Intergovernmental Panel on Climate Change [Masson-Delmotte, V. / Zhai, P. / Pirani, A. / Connors, S. L. / Péan, C. / Berger, S. / Caud, N. / Chen, Y. / Goldfarb, L. / Gomis, M. I. / Huang, M. / Leitzell, K. / Lonnoy, E. / Matthews, J. B. R. / Maycock, T. K. / Waterfield, T. / Yelekçi, O. / Yu, R. / Zhou, B. (Hrsg.)]. Cambridge University Press.
105. Zhang, G. / Murakami, H. / Knutson, T. R. / Mizuta, R. / Yoshida, K.: Tropical cyclone motion in a changing climate. Sci. Adv.6, eaaz7610 (2020).
106. Eckstein, D. / Künzel, V. / Schäfer, L.: Global Climate Risk Index 2021 – Who Suffers Most from Extreme Weather Events? Weather-Related Loss Events in 2019 and 2000–2019. Germanwatch (2021).
107. Sherwood, A. / Bradley, M. / Rossi, L. / Guiam, R. / Mellicke, B.: Resolving Post-Disaster Displacement. Insights from the Philippines after Typhoon Haiyan (Yolanda). Brookings & IOM (2015).
108. Ching, P. K. / de los Reyes, V. C. / Sucaldito, M. N. / Tayag, E.: An assessment of disaster-related mortality post-Haiyan in Tacloban City. Western Pac Surveill Response J.; 6:34–38, veröffentlicht am 6.11.2015; https://doi.org/10.5365/WPSAR.2015.6.2.HYN_005.
109. Walch, C.: Evacuation ahead of natural disasters. Evidence from cyclone Phailin in India and typhoon Haiyan in the Philippines. Geo: Geography and Environment (2018); e00051. https://doi.org/10.1002/geo2.51.
110. https://www.shelterbox.de/.
111. Sherwood, A. et al.: Resolving Post-Disaster Displacement, a. a. O.
112. Vinke, K / Blocher, J. / Becker, M- / Ebay, J. S. / Fong, T. / Kambon, A.: Home lands. Island and archipelagic states' policymaking for human mobility in the context of climate change, (2020).
113. Brewer, L. / Casco, R. / Hills, S. / Kamiya, K. / Kelly, B. / Navldad, C. / Moniz Pereira, M. / Price, J. / Sta Clara, R.: Impact of Livelihood Recovery Initiatives on Reducing Vulnerability to Human Trafficking and Illegal Recruitment. Lessons from Typhoon Haiyan (IOM and ILO, 2015).
114. Ibid.
115. US State Department: Trafficking in Persons Report, (2021).
116. Brashares, J. S. / Abrahms, B. / Fiorella, K. J. / Golden, C. D. / Hojnowski, C. E. / Marsh, R. A. / McCauley, D. J. / Nuñez, T. A. / Seto, K. / Withey, L.: Wildlife decline and social conflict. Science 345, 376–378 (2014).
117. Vinke, K. et al.: Home lands, l.c.

118. Gregorio, N.: The Rise and Fall of Jeepneys in Metro Manila, Philippines. Stanford Future Bay Initiative (2018).
119. IOM Philippines, Climate Change and Sustainability Unit: Framing the Human Narrative of Migration in the Context of Climate Change (2020).
120. Sengupta, S. / Fountain, H.: The Biggest Refugee Camp Braces for Rain: ›This Is Going to Be a Catastrophe‹, New York Times, 14.3.2018.
121. OCHA: Rohingya Refugee Crisis (2020); https://www.unocha.org/rohingya-refugee-crisis.
122. UNHCR, Rohingya refugees restore depleted forest in Bangladesh (2021) https://www.unhcr.org/news/stories/rohingya-refugees-restore-depleted-forest-bangladesh.
123. Crayton, A. et al.: Narratives and Needs. Analyzing Experiences of Cyclone Amphan Using Twitter Discourse (2020).
124. Vinke, K.: Unsettling Settlements-Cities, Migrants, Climate Change. Rural-Urban Climate Migration as Effective Adaptation? LIT Verlag (2019).
125. Mitlin, D. / Satterthwaite, D.: Urban poverty in the global South. Scale and Nature, Routledge (2013); https://www.routledge.com/Urban-Poverty-in-the-Global-South-Scale-and-Nature/Mitlin-Satterthwaite/p/book/9780415624671.
126. Inskip, C. / Fahad, Z. / Tully, R. / Roberts, T. / MacMillan, D.: Understanding carnivore killing behaviour. Exploring the motivations for tiger killing in the Sundarbans, Bangladesh, Biological Conservation, Band 180, S. 42–50 (2014); https://doi.org/10.1016/j.biocon.2014.09.028.
127. UNICEF UK: Futures at Risk – Protecting the rights of children on the move in a changing climate (2021).
128. Ibid & UNICEF: Children uprooted in a changing climate – Turning challenges into opportunities with and for young people on the move (2021).
129. Habekuß, F.: Weltrettung, Versuch Nummer 26, in: Die Zeit, 12.11.2021.
130. WBGU, German Advisory Council on Global Change: Humanity on the move: Unlocking the transformative power of cities. WBGU (2016).
131. Khan, M. R. / Huq, S. / Risha, A. N. / Alam, S. S.: High-density population and displacement in Bangladesh. Science. 372, 1290–1293 (2021).
132. Clement, V. / Rigaud, K. K. / de Sherbinin, A. / Jones, B. / Adamo, S. / Schewe, J. / Sadiq, N. / Shabahat, E.: Groundswell Part 2. Acting on Internal Climate Migration. Washington, D. C., The World Bank (2021).
133. Boulton, C. A. / Lenton, T. M. / Boers, N.: Pronounced loss of Amazon rainforest resilience since the early 2000s. Nat. Clim. Chang. 12, 271–278 (2022); https://doi.org/10.1038/s41558-022-01287-8.
134. Qin, Y. / Xiao, X. / Wigneron, J. P. et al.: Carbon loss from forest degradation exceeds that from deforestation in the Brazilian Amazon. Nat. Clim. Chang. 11, 442–448 (2021); https://doi.org/10.1038/s41558-021-01026-5.
135. Imazon: Deforestation in the Brazilian Amazon (2021); https://imazon.org.br/en/imprensa/deforestation-in-the-brazilian-amazon-reached-2-095-km%C2%B2-in-july-and-the-last-12-months-cumulative-is-the-highest-in-10-years/.
136. Lenton, T. M. et al.: Climate tipping points—too risky to bet against. Nature 575, 592–595 (2019); https://doi.org/10.1038/d41586-019-03595-0.
137. Escobar, H.: There's no doubt that Brazil's fires are linked to deforestation, scientists say, (2019); https://www.science.org/content/article/theres-no-doubt-brazils-fires-are-caused-deforestation-scientists-say.
138. Alessi, G.; Salles vê "oportunidade" com coronavírus para "passar de boiada" desregulação da proteção ao meio ambiente. El Pais Brasil (2020); https://brasil.elpais.com/brasil/2020-05-22/salles-ve-oportunidade-com-coronavirus-para-passar-de-boiada-desregulacao-da-protecao-ao-meio-ambiente.html.
139. Ibid.

140. Leitão, M.: Ódio de Weintraub pelo termo "povos indígenas" contraria a Constituição. Veja (2020); https://veja.abril.com.br/coluna/matheus-leitao/odio-de-weintraub-pelo-termo-povos-indigenas-contraria-a-constituicao.
141. Boadle, A.: Brazil's Bolsonaro blames indigenous people for Amazon fires in U.N. speech, (2020); https://www.reuters.com/article/world/asia-pacific/brazils-bolsonaro-blames-indigenous-people-for-amazon-fires-in-un-speech-idUSKCN26E0AL/.
142. Simões, M.: Brazil's Bolsonaro on the Environment, in His Own Words, New York Times, (2019); https://www.nytimes.com/2019/08/27/world/americas/bolsonaro-brazil-environment.html
143. República Federativa do Brasil, Diário da Camara dos Deputados, (1998); https://imagem.camara.gov.br/Imagem/d/pdf/DCD16ABR1998.pdf#page=33.
144. ›Spiegel‹ (1968): Arsen und Zuckerstückchen; https://www.spiegel.de/politik/arsen-und-zuckerstueckchen-a-23fa92a8-0002-0001-0000-000046093905?context=issue.
145. Cambridge Dictionary: indigenous, Cambridge University Press, (2024); https://dictionary.cambridge.org/dictionary/english/indigenous.
146. Schuster, R. / Germain, R. R. / Bennett, J. R. / Reo, N. J. / Arcese, P.: Vertebrate biodiversity on indigenous-managed lands in Australia, Brazil, and Canada equals that in protected areas (2019), Environmental Science & Policy, Volume 101; https://doi.org/10.1016/j.envsci.2019.07.002.
147. Schleicher, J. / Zaehringer, J. G. / Fastré, C. et al.: Protecting half of the planet could directly affect over one billion people. Nat Sustain 2, 1094–1096 (2019); https://doi.org/10.1038/s41893-019-0423-y.
148. Baragwanath, K. / Bayi, E.: Collective property rights reduce deforestation in the Brazilian Amazon. Proceedings of the National Academy of Sciences 117, 20495–20502 (2020).
149. Salgado, S.: Amazônia, Taschen (2021).
150. IPCC, 2022: 6th Assessment Report, Working Group II, Factsheet Biodiversity – Climate Change Impacts and Risks; https://report.ipcc.ch/ar6wg2/pdf/IPCC_AR6_WGII_FactSheet_Biodiversity.pdf.
151. For those who want to learn more about biodiversity I recommed: Thonicke, K., et al.: 10 Must Knows from Biodiversity Science 2022, (2022); https://zenodo.org/records/6257527#.Ykb20ShBzZs.
152. Radchuk, V. / Reed, T. / Teplitsky, C. et al.: Adaptive responses of animals to climate change are most likely insufficient. Nat Commun 10, 3109 (2019); https://doi.org/10.1038/s41467-019-10924-4.
153. IPCC, 2022: 6th Assessment Report, Working Group II, Factsheet Biodiversity – Climate Change Impacts and Risks.
154. Steffens, D. / Habekuß, F.: Über Leben – Zukunftsfrage Artensterben. Wie wir die Ökokrise überwinden. Penguin (2020).
155. Dinerstein, E. / Joshi, A. R. / Vynne, C. / Lee, A. T. L. / Pharand-Deschênes, F. / França, M. / Fernando, S. / Birch, T. / Burkart, K. / Asner, G. P. / Olson, D.: A »Global Safety Net« to reverse biodiversity loss and stabilize Earth's climate. Sci. Adv.6, eabb2824 (2020).
156. Hirschhausen, E.: Mensch, Erde! Wir könnten es so schön haben, dtv (2021).
157. WHO: Air Pollution (2022); https://www.who.int/health-topics/air-pollution#tab=tab_1.
158. Vohra, K. / Vodonos, A. / Schwartz, J. / Marais, E. A. / Sulprizio, M. P. / Mickley, L. J.: Global mortality from outdoor fine particle pollution generated by fossil fuel combustion. Results from GEOS-Chem, Environmental Research, Band 195 (2021).
159. WHO: Billions of people still breathe unhealthy air (2022); https://www.who.int/news/item/04-04-2022-billions-of-people-still-breathe-unhealthy-air-new-who-data.
160. Romanello, M. et al.: The 2021 report of the Lancet Countdown on health and climate change: code red for a healthy future, The Lancet (2021); https://doi.org/10.1016/S0140-6736(21)01787-6.
161. Schwerdtle, P. N. et al.: Environ. Res. Lett. 15 103006 (2020).
162. Bergmann, J.: At risk of deprivation. The multidimensional well-being impacts of climate migration and immobility in Peru, Springer VS, Wiesbaden (2024).

163. Bergmann, J.: Planned relocation in Peru. Advancing from well-meant legislation to good practice. J Environ Stud Sci 11, 365–375 (2021); https://doi.org/10.1007/s13412-021-00699-w.
164. Ibid.
165. Bergmann, J. / Vinke, K. / Fernández Palomino, C. A. / Gornott, C. / Gleixner, S. / Laudien, R. / Lobanova, A. / Ludescher, J. / Schellnhuber, H. J.: Assessing the Evidence. Climate Change and Migration in Peru. Potsdam Institute for Climate Impact Research (PIK), Potsdam, and International Organization for Migration (IOM), Genf (2021).
166. Mora, C. / Dousset, B. / Caldwell, I. et al.: Global risk of deadly heat. Nature Clim Change 7, 501–506 (2017); https://doi.org/10.1038/nclimate3322.
167. Blocher, J. / Bergmann, J. / Upadhyay, H. / Vinke, K.: Hot, wet and deserted: Climate Change and Internal Displacement in India, Peru, and Tanzania. Insights from the EPICC Project. Background Paper for the Internal Displacement Monitoring Centre (GRID 2021). Internal Displacement Monitoring Centre (IDMC), Genf; https://www.internal-displacement.org/global-report/grid2021/downloads/background_papers/background_paper-climatechange.pdf.
168. Seehaus, T. et al.: »Changes of the Tropical Glaciers Throughout Peru Between 2000 and 2016 – Mass Balance and Area Fluctuations«, The Cryosphere Discussions (2019); https://doi.org/10.5194/tc-2018-289.
169. Blocher, J. et al.: Hot, wet and deserted (GRID 2021).
170. Parsa, S. et al.: Lima 2035, Food System Vision Prize, Rockefeller Foundation (2021); https://challenges.openideo.com/apps/IMT/UploadedFiles/00/f_92dfa7a6e543f07c6672931a76f59f2e/attachments_d9bc35c3-ceea-451b-9b25-77de52867b73.pdf?v=1647829912.
171. More information can be found on the Spanish website: »www.lossinagua.org«.
172. Ludescher, J. / Martin, M. / Boers, N. / Bunde, A. / Ciemer, C. / Fan, J. et al.: Network-Based Forecasting of Climate Phenomena, Proceedings of the National Academy of Sciences, 118.47 (2021); e1922872118 https://doi.org/10.1073/pnas.1922872118.
173. Cai, W. / Borlace, S. / Lengaigne, M. et al.: Increasing frequency of extreme El Niño events due to greenhouse warming. Nature Clim Change 4, 111–116 (2014); https://doi.org/10.1038/nclimate2100 und Cai, W. / Santoso, A. / Collins, M. et al.: Changing El Niño – Southern Oscillation in a warming climate. Nat Rev Earth Environ 2, 628–644 (2021); https://doi.org/10.1038/s43017-021-00199-z.
174. Cai, W. / McPhaden, M. J. / Grimm, A. M. et al.: Climate impacts of the El Niño – Southern Oscillation on South America. Nat Rev Earth Environ 1, 215–231 (2020); https://doi.org/10.1038/s43017-020-0040-3.
175. Davis, M.: Late Victorian Holocausts: El Niño Famines and the Making of the Third World, Verso Books (2001/2017).
176. Oxfam: An Economy for the 99 % (2017).
177. Oxfam: Inequality Kills. The unparalleled action needed to combat unprecedented inequality in the wake of COVID-19 (2022).
178. Takahashi, K. / Martínez, A. G.: The very strong coastal El Niño in 1925 in the far-eastern Pacific. Clim Dyn 52, 7389–7415 (2019); https://doi.org/10.1007/s00382-017-3702-1.
179. Rodríguez-Morata, C. / Díaz, H. F. / Ballesteros-Canovas, J. A. et al.: The anomalous 2017 coastal El Niño event in Peru. Clim Dyn 52, 5605–5622 (2019); https://doi.org/10.1007/s00382-018-4466-y.
180. Reuben, A. / Frischtak, H. / Berky, A. / Ortiz, E. J. / Morales, A. M. / Hsu-Kim, H. et al.: Elevated hair mercury levels are associated with neurodevelopmental deficits in children living near artisanal and small-scale gold mining in Peru. GeoHealth, 4; e2019GH000222. https://doi.org/10.1029/2019GH000222.
181. Martinez, G. / McCord, S. A. / Driscoll, C. T. / Todorova, S. / Wu, S. / Araújo, J. F. / Vega, C. M. / Fernandez, L. E.: Mercury Contamination in Riverine Sediments and Fish Associated with Artisanal and Small-Scale Gold Mining in Madre de Dios, Peru. Int. J. Environ. Res. Public Health 2018, 15, 1584; https://doi.org/10.3390/ijerph15081584.

182. Gilbert, S. G. / Grant-Webster, K. S.: Neurobehavioral effects of developmental methylmercury exposure. Environ Health Perspect 103 (Suppl 6): 135–142 (1995).
183. Yard, E. E. / Horton, J. / Schier, J. G. et al.: Mercury Exposure Among Artisanal Gold Miners in Madre de Dios, Peru: A Cross-sectional Study. J. Med. Toxicol. 8, 441–448 (2012); https://doi.org/10.1007/s13181-012-0252-0.
184. Fadrique, B. / Báez, S. / Duque, Á. et al.: Widespread but heterogeneous responses of Andean forests to climate change. Nature 564, 207–212 (2018); https://doi.org/10.1038/s41586-018-0715-9.
185. Feeley, K. J. / Silman, M. R. / Bush, M. B. / Farfan, W. / Cabrera, K. G. / Malhi, Y. / Meir, P. / Revilla, N. S. / Quisiyupanqui, M. N. R. / Saatchi, S.: Upslope migration of Andean trees (2011). Journal of Biogeography, 38: 783–791; https://doi.org/10.1111/j.1365-2699.2010.02444.x.
186. Kolbert, E.: The Sixth Extinction - An Unnatural History, Macmillan, (2015).
187. Mammoths, too, probably fell victim to a climatic warm phase in the first place. However, their human hunters probably also played a role in their extinction.
188. Xu, C. / Kohler, T. A. / Lenton, T. M. / Svenning, J.-C. / Scheffer, M.: Future of the human climate niche. Proceedings of the National Academy of Sciences 117, 11350–11355 (2020).
189. Energieagentur Rheinland-Pfalz: Gemeinsamer Abschlussbericht 2021. Wärme- und Gasversorgung im Ahrtal nach der Flutkatastrophe (2022).
190. Emcke, C.: Wir können nicht weiterleben wie bisher, 30.7.2021, Süddeutsche Zeitung.
191. Junghänel u. a.: Hydro-klimatologische Einordnung der Stark- und Dauerniederschläge in Teilen Deutschlands im Zusammenhang mit dem Tiefdruckgebiet »Bernd« vom 12. bis 19. Juli 2021, Deutscher Wetterdienst (2021).
192. Timperley, J.: The broken $100-billion promise of climate finance—and how to fix it, Nature 598, 400–402 (2021) https://doi.org/10.1038/d41586-021-02846-3.
193. Deutsche Klimafinanzierung: OECD: Klimafinanzierung knackt 100-Milliarden-Marke, (2024), https://www.deutscheklimafinanzierung.de/blog/2024/06/oecd-klimafinanzierung-knackt-100-milliarden-marke/#:~:text=1.,Dollar%20erreicht%20und%20sogar%20%C3%BCberschritten.
194. Bundesanstalt Technisches Hilfswerk: 1 Million Einsatzstunden – Umzug eines Bereitstellungsraumes, um weiterzuhelfen (2021); https://www.thw.de/SharedDocs/Meldungen/THW-LV-HHMVSH/DE/Einsaetze/2021/08/1408BRumzug.html?idImage=16520364&notFirst=true.
195. Robinson, A. / Lehmann, J. / Barriopedro, D. / Rahmstorf, S. / Coumou, D.: Increasing heat and rainfall extremes now far outside the historical climate. npj climate and atmospheric science (2021).
196. World Weather Attribution: Heavy rainfall which led to severe flooding in Western Europe made more likely by climate change, (2024), https://www.worldweatherattribution.org/heavy-rainfall-which-led-to-severe-flooding-in-western-europe-made-more-likely-by-climate-change/.
197. Hoffmann, P. / Lehmann, J. / Fallah, B. H. / Hattermann, F. F.: Atmosphere similarity patterns in boreal summer show an increase of persistent weather conditions connected to hydro-climatic risks. Scientific Reports (2021).
198. Otto, F.: Angry Weather: Heat Waves, Floods, Storms, and the New Science of Climate Change. Greystone Books (2023).
199. CIA World Factbook (2021): Electricity-Exports (2016); https://www.cia.gov/the-world-factbook/field/electricity-exports/country-comparison.
200. Robine, J.-M. / Cheung, S. L. K. / Le Roy, Setal, S. et al.: Death toll exceeded 70,000 in Europe during the summer Of 2003. C R Biol 331:171–178 (2008).
201. Winklmayr C. / Muthers S. / Niemann, H. / Mücke, H. G. / an der Heiden, M.: Heat-related mortality in Germany from 1992 to 2021. Deutsches Ärzteblatt Int 2022; 119: 451–7; https://doi.org/10.3238/arztebl.m2022.0202.
202. https://www.klimahaus-bremerhaven.de/.

203. Steinbach, A. / Ty, M. : Das letzte Eis, Klimahaus Bremerhaven, Reisedepeschen (2022).
204. Plöger, S. / Schlenker, R.: Die Alpen und wie sie unser Wetter beeinflussen. Malik (2022).
205. Zemp, M. / Haeberli, W. / Hoelzle, M. / Paul, F.: Alpine glaciers to disappear within decades (2006)? Geophys. Res. Lett., 33, L13504, https://doi.org/10.1029/2006GL026319.
206. Alois (Wisi) Infanger founded the successful mountain tour agency Montanara in the early 90s: www.montanara.ch.
207. Ty, M. / Steingässer, J.: Nordsee Südsee – Zwei Welten im Wandel, Knesebeck (2020).
208. Ministerium für Energiewende, Landwirtschaft, Umwelt und ländliche Räume des Landes Schleswig-Holstein (2015): Strategie für das Wattenmeer 2100.
209. Parenti, C.: Die Welt in einem Laib Brot (2011), Le Monde diplomatique; https://monde-diplomatique.de/artikel/!245356.
210. On the Youtube channel of the German Climate Foundation, the living conditions in the Samos camp are documented: "Climate Escape Documentary: https://www.youtube.com/watch?v=zVDw6AEskz8.
211. The Samos Volunteers have provided information about the Vathy Camp on their website: https://www.samosvolunteers.org/.
212. Christides, G. / Popp, M.: Wie Europa das Recht bricht (2020).
213. United Nations: Secretary-General Warns of Climate Emergency, Calling Intergovernmental Panel's Report ›a File of Shame‹, While Saying Leaders ›Are Lying‹, Fuelling Flames (2022); https://www.un.org/press/en/2022/sgsm21228.doc.htm.
214. Horton, R. M. / Sherbinin, A. / de Wrathall, D. / Oppenheimer, M.: Assessing human habitability and migration (2021). Science 372, 1279–1283; https://doi.org/10.1126/science.abi8603.
215. Harper, A. / Vinke, K.: Climate Change and the Future of Safe Returns (2020), UNHCR & PIK; https://www.unhcr.org/5fb28b504.pdf.
216. World Meteorological Organization: WMO update: 50:50 chance of global temperature temporarily reaching 1.5°C threshold in next five years, WMO & UK Met Office (2022); https://public.wmo.int/en/media/press-release/wmo-update-5050-chance-of-global-temperature-temporarily-reaching-15%C2%B0c-threshold.
217. https://climate.copernicus.eu/new-record-daily-global-average-temperature-reached-july-2024
218. Schulte, Hildegard: Zeitzeichen – 27.06.1973 – Todestag von Odd Nansen, WDR (2018); https://www1.wdr.de/radio/wdr5/sendungen/zeitzeichen/odd-nansen-100.html.
219. Ibid.
220. The Nansen Initiative: Agenda for the Protection of Cross-Border Displaced Persons in the Context of Disasters and Climate Change (2015); https://www.eda.admin.ch/dam/eda/en/documents/aussenpolitik/menschenrechte-menschliche-sicherheit/2017-und-aelter/Nansen-Initiative-Schutzagenda-Volume-1_EN.pdf.
221. Warner, K.: Risk, Climate Change and Human Mobility in International Policy. In: Preuß, H. J. / Beier, C. / Messner, D. (Hrsg.): Forced Displacement and Migration. Springer (2022); https://doi.org/10.1007/978-3-658-32902-0_9.
222. Piguet, E. / Kaenzig, R. / Guélat, J.: The uneven geography of research on »environmental migration«. Popul Environ 39, 357–383 (2018); https://doi.org/10.1007/s11111-018-0296-4.
223. UNHCR: Strategic Framework for Climate Action (2021); https://www.unhcr.org/604a26d84.pdf.
224. https://www.uno-fluechtlingshilfe.de/informieren/fluchtursachen/klimawandel.
225. United Nations: Global Compact for Safe, Orderly and Regular Migration (A/RES/73/195 (2018).))
226. The White House: Executive Order on Rebuilding and Enhancing Programs to Resettle Refugees and Planning for the Impact of Climate Change on Migration (2021); https://www.whitehouse.gov/briefing-room/presidential-actions/2021/02/04/executive-order-on-rebuilding-and-enhancing-programs-to-resettle-refugees-and-planning-for-the-impact-of-climate-change-on-migration/.

227. The White House: Report on the Impact of Climate Change on Migration (2021); https://www.whitehouse.gov/wp-content/uploads/2021/10/Report-on-the-Impact-of-Climate-Change-on-Migration.pdf.
228. Fachkommission Fluchtursachen: Krisen vorbeugen, Perspektiven schaffen, Menschen schützen – Bericht der Fachkommission Fluchtursachen der Bundesregierung (2021).
229. WBGU – Wissenschaftlicher Beirat der Bundesregierung Globale Umweltveränderungen: Zeitgerechte Klimapolitik: Vier Initiativen für Fairness. Politikpapier 9. WBGU (2018).
230. Ibid.
231. Vinke, K. / Donges, J. / Gardiner, S. / Gärtner, J. / Thornton, F. / Schellnhuber, H. J.: The Freedom to Move in Response to Uninhabitability: Enabling Climate Migration by a Nansen-Type Passport (in Vorbereitung).
232. Luscombe, R.: Trump's border wall reportedly in severe disrepair in Arizona. The Guardian (2021).
233. Roth, C. et al.: Klimabedingte Migration, Flucht und Vertreibung – Eine Frage globaler Gerechtigkeit. Deutscher Bundestag, 19. Wahlperiode, Drucksache 19/ 15781 (2019); https://dserver.bundestag.de/btd/19/157/1915781.pdf.
234. Deutscher Bundestag: Anträge zur »globalen Klimagerechtigkeit« (2019); https://www.bundestag.de/dokumente/textarchiv/2019/kw50-de-migration-klima-670582.
235. Dernbach, A.: Welche Lehren aus der Kölner Silvesternacht gezogen wurden. Der Tagesspiegel (2020).
236. Deutscher Bundestag, Mediathek: Globale Klimagerechtigkeit. Beratung des Antrags der Fraktion Bündnis 90/Die Grünen. Klimabedingte Migration, Flucht und Vertreibung – Eine Frage globaler Gerechtigkeit (2019); https://www.bundestag.de/mediathek?videoid=7407501#url=L21lZGlhdGhla292ZXJsYXk/dmlkZW9pZD03NDA3NTAx&mod=mediathek.
237. Hickel, J.: Quantifying national responsibility for climate breakdown: an equality-based attribution approach for carbon dioxide emissions in excess of the planetary boundary (2020), The Lancet Planetary Health, Volume 4, Issue 9; https://doi.org/10.1016/S2542-5196(20)30196-0.
238. Brand, U. / Wissen, M.: Imperiale Lebensweise – Zur Ausbeutung von Mensch und Natur im globalen Kapitalismus. Oekom (2017).
239. Gardener, B. / Yeampierre, E.: Unequal Impact. The Deep Links Between Racism and Climate Change. Yale School of the Environment (2020).
240. Adeola, F. O. / Picou, J. S.: Race, social capital, and the health impacts of Katrina: Evidence from the Louisiana and Mississippi Gulf Coast. Human Ecology Review 19(1):10–24 (2012).
241. National Academies of Sciences, Engineering, and Medicine: Framing the Challenge of Urban Flooding in the United States. The National Academies Press (2019); https://doi.org/10.17226/25381.
242. Ibid.
243. Bullard, R. / Wright, B.: Race, Place, and Environmental Justice After Hurricane Katrina – Struggles to Reclaim, Rebuild, and Revitalize New Orleans and the Gulf Coast. Routledge (2009).
244. Li, W. / Airriess, C. A. / Chen, A. C.-C. / Leong, K. J. / Keith, V.: Katrina and Migration: Evacuation and Return by African Americans and Vietnamese Americans in an Eastern New Orleans Suburb. The Professional Geographer, 62:1, 103–118 (2010), https://doi.org/10.1080/00330120903404934.
245. Salas, R. N.: Environmental Racism and Climate Change—Missed Diagnoses. New England Journal of Medicine (2021), https://doi.org/10.1056/NEJMp2109160.
246. Hoffman, J. S. / Shandas, V. / Pendleton, N.: The Effects of Historical Housing Policies on Resident Exposure to Intra-Urban Heat: A Study of 108 US Urban Areas. Climate 2020, 8, 12; https://doi.org/10.3390/cli8010012.
247. Ibid.
248. Ituen, I. / Tatu Hey, L.: Room – Environmental Racism in Germany Studies, knowledge gaps, and their relevance to environmental and climate justice, Heinrich Böll Stiftung (2021).

249. Hasters, A.: Was weiße Menschen nicht über Rassismus hören wollen, aber wissen sollten. hanserblau (2019).
250. Ibid.
251. IPCC, 2022: Climate Change 2022: Impacts, Adaptation, and Vulnerability. Contribution of Working Group II to the Sixth Assessment Report of the Intergovernmental Panel on Climate Change [Pörtner, H.-O. / Roberts, D. C. / Tignor, M. / Poloczanska, E. S. / Mintenbeck, K. / Alegría, A. / Craig, M. / Langsdorf, S. / Löschke, S. / Möller, V. / Okem, A. / Rama, B. (Hrsg.)]. Cambridge University Press.
252. Ibid., Kapitel 7, Box 7.4.
253. Ibid., Kapitel 7.
254. Oxfam: Time to care – Unpaid and underpaid care work and the global inequality crisis (2020).
255. UNDP: 2022 Special Report – New threats to human security in the Anthropocene – Demanding greater solidarity. United Nations Development Programme (2022).
256. Ibid.
257. Belmin, C. / Hoffmann, R. / Pichler, P. P. / Weisz, H.: Fertility transition powered by women's access to electricity and modern cooking fuels. Nat Sustain 5, 245–253 (2022); https://doi.org/10.1038/s41893-021-00830-3.
258. Schuster, R. / Wachter, K. / Hussain, F. / Gartin, M.: Gendered effects of climate change and health inequities among forcibly displaced populations: Displaced Rohingya women foster resilience through technology. The Journal of Climate Change and Health, (2024); https://doi.org/10.1016/j.joclim.2024.100303.
259. UNDP: Human Development Report 2020 – The next frontier Human development and the Anthropocene (2020).
260. Schipper, E. L. F. / Ensor, J. / Mukherji, A. / Mirzabaev, A. / Fraser, A. / Harvey, B. / Totin, E. / Garschagen, M. / Pathak, M. / Antwi-Agyei, P. / Tanner, T. / Shawoo, Z.: Equity in climate scholarship: a manifesto for action, Climate and Development, 13:10, 853–856 (2021); https://doi.org/10.1080/17565529.2021.1923308.
261. Forschung & Lehre: Frauenanteil bei Professuren stagniert (2021); https://www.forschung-und-lehre.de/politik/frauenanteil-bei-professuren-stagniert-4084.
262. Lunz, K.: Die Zukunft der Außenpolitik ist feministisch – Wie globale Krisen gelöst werden müssen. Ullstein (2022).
263. Supran, G. / Oreskes, N.: Environmental Research Letters. 12 084019 (2017).
264. Rahmstorf, S.: Ein Forscher sagte schon 1977 den Klimawandel voraus – leider arbeitete er bei Exxon. Der Spiegel (2019).
265. Black-Kalinsky, C.: My father warned Exxon about climate change in the 1970s. They didn't listen. The Guardian (2016); https://www.theguardian.com/commentisfree/2016/may/25/exxon-climate-change-greenhouse-gasses.
266. World Economic Forum (2021): Global Gender Gap Report 2021.
267. Carrington, D.: Climate change denial is evil, says Mary Robinson. The Guardian (2019); https://www.theguardian.com/environment/2019/mar/26/climate-change-denial-is-evil-says-mary-robinson.
268. The concept of a "World Citizen Movement" originates from a special report of the WBGU (2014): Climate Protection as a World Citizen Movement.
269. Riise J. / Adeyemi, K.: Case study: Makoko floating school. Curr Opin Environ Sustain 13:58–60 (2015).
270. Barnes, B. / Cao, H. / Drab, T. / Pearson, J.: Design of sustainable relief housing in Ethiopia: An implementation of cradle to cradle design in earthbag construction. Am J Environ Sci 5(2):137–144 (2009).
271. Horton, B. P. et al.: Estimating global mean sea-level rise and its uncertainties by 2100 and 2300 from an expert survey. npj Climate and Atmospheric Science 3 (2020). https://doi.org/10.1038/s41612-020-0121-5.

272. Kulp, S.A. / Strauss, B.H.: New elevation data triple estimates of global vulnerability to sea-level rise and coastal flooding. Nat Commun 10, 4844 (2019); https://doi.org/10.1038/s41467-019-12808-z.
273. Greiner, T.: Architektur für den Klimawandel – Die Welt lernt schwimmen. Süddeutsche Zeitung (2015); https://www.sueddeutsche.de/wissen/architektur-fuer-den-klimawandel-die-welt-lernt-schwimmen-1.2544091.
274. Roehrl, A. / Aufmkolk, T.: Zukunft des Wohnens – Schwimmende Häuser. SWR Planet Wissen (2019).
275. UN-Habitat: UN-Habitat and partners unveil OCEANIX Busan, the world's first prototype floating city (2022).
276. Ingels, B. / Sundlin, D.: Oceanix City. BIG-Bjarke Ingels (2022); https://big.dk/#projects-sfc.
277. See for example »Morgenstadt – Stadt der Zukunft Initiative«, Fraunhofer Gesellschaft (2015).
278. Churkina, G. / Organschi, A. / Reyer, C.P.O. et al.: Buildings as a global carbon sink. Nature Sustainability 3, 269–276 (2020); https://doi.org/10.1038/s41893-019-0462-4.
279. Von der Leyen, U.: Ein Neues Europäisches Bauhaus, FAZ 20.2.2021.
280. World Weather Attribution: Climate Change made devastating early heat in India and Pakistan 30 times more likely (2022); https://www.worldweatherattribution.org/climate-change-made-devastating-early-heat-in-india-and-pakistan-30-times-more-likely/.
281. The video of the discussion can be viewed here: https://www.youtube.com/watch?v=rJtoXmCh0Ns.
282. Harvey, F.: Climate experts in dismay at choice of Mathias Cormann as OECD chief. The Guardian (2021); https://www.theguardian.com/business/2021/mar/12/climate-experts-in-dismay-at-choice-of-mathias-cormann-as-oecd-chief.
283. https://globalsolaratlas.info/ and https://globalwindatlas.info/.

The manufacturer's authorised representative in the EU is Springer Nature Customer Service Centre GmbH, Europaplatz 3, 69115 Heidelberg, Germany. If you have any concerns regarding our products, please contact ProductSafety@springernature.com

Printed and bound by CPI Group (UK) Ltd, Croydon, CR0 4YY

26/03/2026

02078941-0011